**To Roger Penrose**

For his beautiful, surprising discoveries in mathematics,
physics and cosmology; for his deep creative insight
into how the universe operates; and for his humility in
not supposing that he is exploring only the products of
human minds.

## SPECTRUM SERIES

The Spectrum Series of the Mathematical Association of America was so named to reflect its purpose: to publish a broad range of books including biographies, accessible expositions of old or new mathematical ideas, reprints and revisions of excellent out-of-print books, popular works, and other monographs of high interest that will appeal to a broad range of readers, including students and teachers of mathematics, mathematical amateurs, and researchers.

*All the Math That's Fit to Print,* by Keith Devlin
*Circles: A Mathematical View,* by Dan Pedoe
*Complex Numbers and Geometry,* by Liang-shin Hahn
*Cryptology,* by Albrecht Beutelspacher
*Five Hundred Mathematical Challenges,* Edward J. Barbeau, Murray S. Klamkin, and William O. J. Moser
*From Zero to Infinity,* by Constance Reid
*I Want to be a Mathematician,* by Paul R. Halmos
*Journey into Geometries,* by Marta Sved
*JULIA: a life in mathematics,* by Constance Reid
*The Last Problem,* by E. T. Bell (revised and updated by Underwood Dudley)
*The Lighter Side of Mathematics: Proceedings of the Eugène Strens Memorial Conference on Recreational Mathematics & its History,* edited by Richard K. Guy and Robert E. Woodrow
*Lure of the Integers,* by Joe Roberts
*Mathematical Carnival,* by Martin Gardner
*Mathematical Circus,* by Martin Gardner
*Mathematical Cranks,* by Underwood Dudley
*Mathematical Magic Show,* by Martin Gardner
*Mathematics: Queen and Servant of Science,* by E. T. Bell
*Memorabilia Mathematica,* by Robert Edouard Moritz
*New Mathematical Diversions,* by Martin Gardner
*Numerical Methods that Work,* by Forman Acton
*Out of the Mouths of Mathematicians,* by Rosemary Schmalz
*Penrose Tiles to Trapdoor Ciphers . . . and the Return of Dr. Matrix,* by Martin Gardner
*Polyominoes,* by George Martin
*The Search for E. T. Bell, also known as John Taine,* by Constance Reid
*Shaping Space,* edited by Marjorie Senechal and George Fleck
*Student Research Projects in Calculus,* by Marcus Cohen, Edward D. Gaughan, Arthur Knoebel, Douglas S. Kurtz, and David Pengelley
*The Trisectors,* by Underwood Dudley
*The Words of Mathematics,* by Steven Schwartzman

MAA Service Center
P. O. Box 91112
Washington, DC 20090-1112
800-331-1MAA    FAX 202-206-9789

# *Contents*

# Preface

This book is a collection of *Scientific American* columns that I wrote over a period of 25 years. It is the thirteenth such collection. The unifying theme, if there must be one, is recreational mathematics; that is, mathematics presented in a spirit of play. As with earlier books, the columns have been corrected and expanded with addendums based on current feedback from readers. So much has happened since to Penrose tilings (especially their unexpected applications to crystal theory), to public key cryptosystems and to the French *Oulipo*, that I have written brand new chapters for each of these topics. Also published here for the first time is astonishing news concerning my old and dear friend Dr. Matrix. I report my discovery that Dr. Matrix had not been killed by a Russian KGB agent, as supposed, but is alive and well in Casablanca.

*Martin Gardner*

# CHAPTER 1

# Penrose Tiling

At the end of a 1975 *Scientific American* column on tiling the plane periodically with congruent convex polygons (reprinted in my *Time Travel and Other Mathematical Bewilderments*) I promised a later column on nonperiodic tiling. This chapter reprints my fulfillment of that promise—a 1977 column that reported for the first time a remarkable nonperiodic tiling discovered by Roger Penrose, the noted British mathematical physicist and cosmologist. First, let me give some definitions and background.

A periodic tiling is one on which you can outline a region that tiles the plane by translation, that is, by shifting the position of the region without rotating or reflecting it. M. C. Escher, the Dutch artist, is famous for his many pictures of periodic tilings with shapes that resemble living things. Figure 1 is typical. An adjacent black and white bird constitute a fundamental region that tiles by translation. Think of the plane as being covered with transparent paper on which each tile is outlined. Only if the tiling is periodic can you shift the paper, without rotation, to a new position where all outlines again exactly fit.

An infinity of shapes—for instance the regular hexagon—tile only periodically. An infinity of other shapes tile both periodically and nonperiodically. A checkerboard is easily converted to a nonperiodic tiling by identical isosceles right triangles or by quadrilaterals. Simply bisect

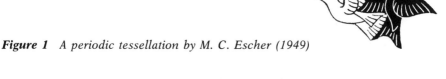

**Figure 1**   *A periodic tessellation by M. C. Escher (1949)*

each square as shown in Figure 2A, left, altering the orientations to prevent periodicity. It is also easy to tile nonperiodically with dominoes.

Isoceles triangles also tile in the radial fashion shown in the center of Figure 2A. Although the tiling is highly ordered, it is obviously not periodic. As Michael Goldberg pointed out in a 1955 paper titled "Central Tessellations," such a tiling can be sliced in half, and then the half planes can be shifted one step or more to make a spiral form of nonperiodic tiling, as shown in Figure 2A, right. The triangle can be distorted in an infinity of ways by replacing its two equal sides with congruent lines, as shown at the left in Figure 2B. If the new sides have straight edges, the result is a polygon of 5, 7, 9, 11 . . . edges that tiles spirally. Figure 3 shows a striking pattern obtained in this way from a nine-sided polygon. It was first found by Heinz Voderberg in a complicated procedure. Goldberg's method of obtaining it makes it almost trivial.

In all known cases of nonperiodic tiling by congruent figures the figure also tiles periodically. Figure 2B, right, shows how two of the Voderberg enneagons go together to make an octagon that tiles periodically in an obvious way.

Another kind of nonperiodic tiling is obtained by tiles that group together to form larger replicas of themselves. Solomon W. Golomb calls them "reptiles." (See Chapter 19 of my book *Unexpected Hanging*.) Figure 4 shows how a shape called the "sphinx" tiles nonperiodically by giving rise to ever larger sphinxes. Again, two sphinxes (with one sphinx rotated 180 degrees) tile periodically in an obvious way.

Are there sets of tiles that tile only nonperiodically? By "only" we mean that neither a single shape or subset nor the entire set tiles periodically, but that by using all of them a nonperiodic tiling is possible. Rotating and reflecting tiles are allowed.

For many decades experts believed no such set exists, but the supposition proved to be untrue. In 1961 Hao Wang became interested in tiling the plane with sets of unit squares whose edges were colored in various ways. They are called Wang dominoes, and Wang wrote a splendid article about them for *Scientific American* in 1965. Wang's problem was to find a procedure for deciding whether any given set of dominoes will tile by placing them so that abutting edges are the same color. Rotations and reflections are not allowed. The problem is important because it relates to decision questions in symbolic logic. Wang conjectured that any set of tiles which can tile the plane can tile it periodically and showed that if this is the case, there is a decision procedure for such tiling.

In 1964 Robert Berger, in his thesis for a doctorate from Harvard University in applied mathematics, showed that Wang's conjecture is

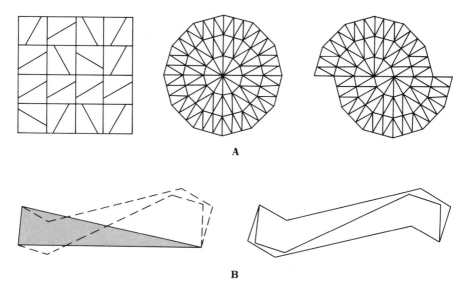

**Figure 2** *(A) Nonperiodic tiling with congruent shapes (B) An enneagon (dotted at left) and a pair of enneagons (right) forming an octagon that tiles periodically*

**Figure 3**   *A spiral tiling by Heinz Voderberg*

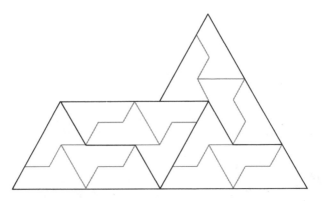

**Figure 4**   *Three generations of sphinxes in a nonperiodic tiling*

false. There is no general procedure. Therefore there is a set of Wang dominoes that tiles only nonperiodically. Berger constructed such a set, using more than 20,000 dominoes. Later he found a much smaller set of 104, and Donald Knuth was able to reduce the number to 92.

It is easy to change such a set of Wang dominoes into polygonal tiles that tile only nonperiodically. You simply put projections and slots on the edges to make jigsaw pieces that fit in the manner formerly prescribed by colors. An edge formerly one color fits only another formerly the same color, and a similar relation obtains for the other colors. By allowing such tiles to rotate and reflect Robinson constructed six tiles (see Figure 5) that force nonperiodicity in the sense explained above. In 1977 Robert Ammann found a different set of six tiles that also force nonperiodicity. Whether tiles of this square type can be reduced to less than six is not known, though there are strong grounds for believing six to be the minimum.

At the University of Oxford, where he is Rouse Ball Professor of Mathematics, Penrose found small sets of tiles, not of the square type,

**Figure 5** *Raphael M. Robinson's six tiles that force a nonperiodic tiling*

that force nonperiodicity. Although most of his work is in relativity theory and quantum mechanics, he continues the active interest in recreational mathematics he shared with his geneticist father, the late L. S. Penrose. (They are the inventors of the famous "Penrose staircase" that goes round and round without getting higher; Escher depicted it in his lithograph "Ascending and Descending.") In 1973 Penrose found a set of six tiles that force nonperiodicity. In 1974 he found a way to reduce them to four. Soon afterward he lowered them to two.

Because the tiles lend themselves to commercial puzzles, Penrose was reluctant to disclose them until he had applied for patents in the United Kingdom, the United States and Japan. The patents are now in force. I am equally indebted to John Horton Conway for many of the results of his study of the Penrose tiles.

The shapes of a pair of Penrose tiles can vary, but the most interesting pair have shapes that Conway calls "darts" and "kites." Figure 6A shows how they are derived from a rhombus with angles of 72 and 108 degrees. Divide the long diagonal in the familiar golden ratio of (1 +

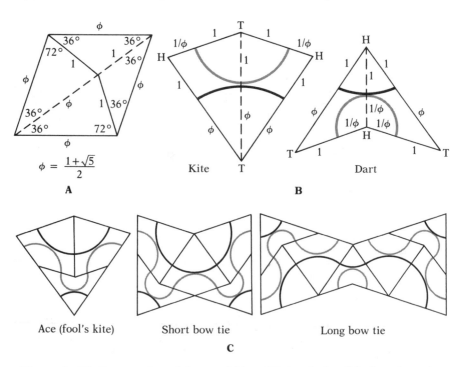

**Figure 6** *(A) Construction of dart and kite (B) A coloring (black and gray) of dart and kite to force nonperiodicity (C) Aces and bow ties that speed constructions*

$\sqrt{5})/2 = 1.61803398 \ldots$ , then join the point to the obtuse corners. That is all. Let phi stand for the golden ratio. Each line segment is either 1 or phi as indicated. The smallest angle is 36 degrees, and the other angles are multiples of it.

The rhombus of course tiles periodically, but we are not allowed to join the pieces in this manner. Forbidden ways of joining sides of equal length can be enforced by bumps and dents, but there are simpler ways. For example, we can label the corners $H$ and $T$ (heads and tails) as is shown in Figure 6B, and then give the rule that in fitting edges only corners of the same letter may meet. Dots of two colors could be placed in the corners to aid in conforming to this rule, but a prettier method, proposed by Conway, is to draw circular arcs of two colors on each tile, shown in the illustration as black and gray. Each arc cuts the sides as well as the axis of symmetry in the golden ratio. Our rule is that abutting edges must join arcs of the same color.

To appreciate the full beauty and mystery of Penrose tiling one should make at least 100 kites and 60 darts. The pieces need be colored on one side only. The number of pieces of the two shapes are (like their areas) in the golden ratio. You might suppose you need more of the smaller darts, but it is the other way around. You need 1.618 . . . as many kites as darts. In an infinite tiling this proportion is exact. The irrationality of the ratio underlies a proof by Penrose that the tiling is nonperiodic because if it were periodic, the ratio clearly would have to be rational.

A good plan is to draw as many darts and kites as you can on one sheet, with a ratio of about five kites to three darts, using a thin line for the curves. The sheet can be photocopied many times. The curves can then be colored with, say, red and green felt-tip pens. Conway has found that it speeds constructions and keeps patterns stabler if you make many copies of the three larger shapes as is shown in Figure 6C. As you expand a pattern, you can continually replace darts and kites with aces and bow ties. Actually an infinity of arbitrarily large *pairs* of shapes, made up of darts and kites, will serve for tiling any infinite pattern.

A Penrose pattern is made by starting with darts and kites around one vertex and then expanding radially. Each time you add a piece to an edge, you must choose between a dart and a kite. Sometimes the choice is forced, sometimes it is not. Sometimes either piece fits, but later you may encounter a contradiction (a spot where no piece can be legally added) and be forced to go back and make the other choice. It is a good plan to go around a boundary, placing all the forced pieces first. They cannot lead to a contradiction. You can then experiment with unforced pieces. It is always possible to continue forever. The more you play with the pieces, the more you will become aware of "forcing rules" that increase efficiency. For example, a dart forces two kites in its concavity, creating the ubiquitous ace.

There are many ways to prove that the number of Penrose tilings is uncountable, just as the number of points on a line is. These proofs rest on a surprising phenomenon discovered by Penrose. Conway calls it "inflation" and "deflation." Figure 7 shows the beginning of inflation. Imagine that every dart is cut in half and then all short edges of the original pieces are glued together. The result: a new tiling (shown in heavy black lines) by larger darts and kites.

Inflation can be continued to infinity, with each new "generation" of pieces larger than the last. Note that the second-generation kite, although it is the same size and shape as a first-generation ace, is formed differently. For this reason the ace is also called a fool's kite. It should never be mistaken for a second-generation kite. Deflation is the same process carried the other way. On every Penrose tiling we can draw smaller and smaller generations of darts and kites. This pattern too goes to infinity, creating a structure that is a fractal (see Chapter 3).

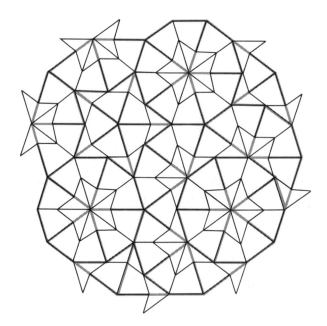

**Figure 7** *How a pattern is inflated*

Conway's proof of the uncountability of Penrose patterns (Penrose had earlier proved it in a different way) can be outlined as follows. On the kite label one side of the axis of symmetry $L$, the other $R$ (for left and right). Do the same on the dart, using $l$ and $r$. Now pick a random point on the tiling. Record the letter that gives its location on the tile. Inflate the pattern one step, note the location of the same point in a second-generation tile and again record the letter. Continuing through higher inflations, you generate an infinite sequence of symbols that is a unique labeling of the original pattern seen, so to speak, from the selected point.

Pick another point on the original pattern. The procedure may give a sequence that starts differently, but it will reach a letter beyond which it agrees to infinity with the former sequence. If there is no such agreement beyond a certain point, the two sequences label distinct patterns. Not all possible sequences of the four symbols can be produced this way, but those that label different patterns can be shown to correspond in number with the number of points on a line.

We have omitted the colored curves on our pictures of tilings because they make it difficult to see the tiles. If you work with colored tiles, however, you will be struck by the beautiful designs created by these curves. Penrose and Conway independently proved that whenever a curve closes, it has a pentagonal symmetry, and the entire region within the curve has a fivefold symmetry. At the most a pattern can have two curves of each color that do not close. In most patterns all curves close.

Although it is possible to construct Penrose patterns with a high degree of symmetry (an infinity of patterns have bilateral symmetry), most patterns, like the universe, are a mystifying mixture of order and unexpected deviations from order. As the patterns expand, they seem to be always striving to repeat themselves but never quite managing it. G. K. Chesterton once suggested that an extraterrestrial being, observing how many features of a human body are duplicated on the left and the right, would reasonably deduce that we have a heart on each side. The world, he said, "looks just a little more mathematical and regular than it is; its exactitude is obvious, but its inexactitude is hidden; its wildness lies in wait." Everywhere there is a "silent swerving from accuracy by an inch that is the uncanny element in everything . . . a sort of secret treason in the universe." The passage is a nice description of Penrose's planar worlds.

There is something even more surprising about Penrose universes. In a curious finite sense, given by the "local isomorphism theorem," all Penrose patterns are alike. Penrose was able to show that every finite region in any pattern is contained somewhere inside every other pattern. Moreover, it appears infinitely many times in every pattern.

To understand how crazy this situation is, imagine you are living on an infinite plane tessellated by one tiling of the uncountable infinity of Penrose tilings. You can examine your pattern, piece by piece, in ever expanding areas. No matter how much of it you explore you can never determine which tiling you are on. It is no help to travel far out and examine disconnected regions, because all the regions belong to one large finite region that is exactly duplicated infinitely many times on all patterns. Of course, this is trivially true of any periodic tessellation, but Penrose universes are not periodic. They differ from one another in infinitely many ways, and yet it is only at the unobtainable limit that one can be distinguished from another.

Suppose you have explored a circular region of diameter $d$. Call it the "town" where you live. Suddenly you are transported to a randomly chosen parallel Penrose world. How far are you from a circular region that exactly matches the streets of your home town? Conway answers with a truly remarkable theorem. The distance from the perimeter of the home town to the perimeter of the duplicate town is never more than $d$ times half of the cube of the golden ratio, or 2.11+ times $d$. (This is an upper bound, not an average.) If you walk in the right direction, you need not go more than that distance to find yourself inside an exact copy of your home town. The theorem also applies to the universe in which you live. Every large circular pattern (there is an infinity of different ones) can be reached by walking a distance in some direction that is certainly less than about twice the diameter of the pattern and more likely about the same distance as the diameter.

The theorem is quite unexpected. Consider an analogous isomorphism exhibited by a sequence of unpatterned digits such as pi. If you pick a finite sequence of 10 digits and then start from a random spot in pi, you are pretty sure to encounter the same sequence if you move far enough along pi, but the distance you must go has no known upper bound, and the expected distance is enormously longer than 10 digits. The longer the finite sequence is, the farther you can expect to walk to find it again. On a Penrose pattern you are always very close to a duplicate of home.

There are just seven ways that darts and kites will fit around a vertex. Let us consider first, using Conway's nomenclature, the two ways with pentagonal symmetry.

The sun (shown in white in Figure 8) does not force the placing of any other piece around it. If you add pieces so that pentagonal symmetry

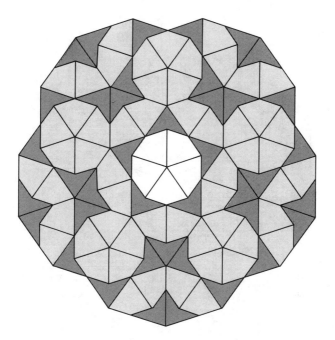

**Figure 8**   *The infinite sun pattern*

is always preserved, however, you will be forced to construct the beautiful pattern shown. It is uniquely determined to infinity.

The star, shown in white in Figure 9, forces the 10 light gray kites around it. Enlarge this pattern, always preserving the fivefold symmetry, and you will create another flowery design that is infinite and unique. The star and sun patterns are the only Penrose universes with perfect pentagonal symmetry, and there is a lovely sense in which they are equivalent. Inflate or deflate either of the patterns and you get the other.

The ace is a third way to tile around a vertex. It forces no more pieces. The deuce, the jack and the queen are shown in white in Figure 10, surrounded by the tiles they immediately force. As Penrose discovered (it was later found independently by Clive Bach), some of the seven vertex figures force the placing of tiles that are not joined to the immediately forced region. Plate 1 shows in deep color the central portion of the king's "empire." (The king is the dark gray area.) All the deep colored tiles are forced by the king. (Two aces, just outside the left and right borders, are also forced but are not shown.)

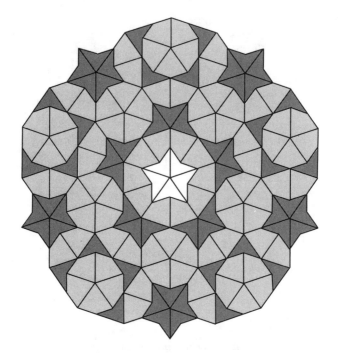

**Figure 9**   *The infinite star pattern*

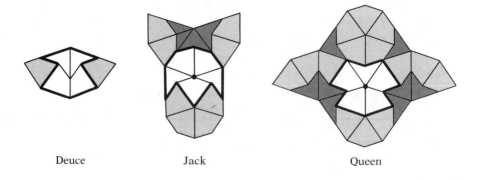

| Deuce | Jack | Queen |

**Figure 10**   *The "empires" of deuce, jack and queen*

This picture of the king's empire was drawn by a computer program written by Eric Regener of Concordia University in Montreal. His program deflates any Penrose pattern any number of steps. The heavy black lines show the domain immediately forced by the king. The thin black lines are a third-generation deflation in which the king and almost all of his empire are replicated.

The most extraordinary of all Penrose universes, essential for understanding the tiles, is the infinite cartwheel pattern, the center of which is shown in Figure 11. The regular decagon at the center, outlined in heavy black (each side is a pair of long and short edges), is what Conway calls a "cartwheel." Every point on any pattern is inside a cartwheel exactly like

**Figure 11** *The cartwheel pattern surrounding Batman*

this one. By one-step inflation we see that every point will be inside a larger cartwheel. Similarly, every point is inside a cartwheel of every generation, although the wheels need not be concentric.

Note the 10 light gray spokes that radiate to infinity. Conway calls them "worms." They are made of long and short bow ties, the number of long ones being in the golden ratio to the number of short ones. Every Penrose universe contains an infinite number of arbitrarily long worms. Inflate or deflate a worm and you get another worm along the same axis. Observe that two full worms extend across the central cartwheel in the infinite cartwheel pattern. (Inside it they are not gray.) The remaining spokes are half-infinite worms. Aside from spokes and the interior of the central cartwheel, the pattern has perfect tenfold symmetry. Between any two spokes we see an alternating display of increasingly large portions of the sun and star patterns.

Any spoke of the infinite cartwheel pattern can be turned side to side (or, what amounts to the same thing, each of its bow ties can be rotated end for end), and the spoke will still fit all surrounding tiles except for those inside the central cartwheel. There are 10 spokes; thus there are $2^{10} = 1024$ combinations of states. After eliminating rotations and reflections, however, there are only 62 distinct combinations. Each combination leaves inside the cartwheel a region that Conway has named a "decapod."

Decapods are made up of 10 identical isosceles triangles with the shapes of enlarged half darts. The decapods with maximum symmetry are the buzzsaw and the starfish shown in Figure 12. Like a worm, each triangle can be turned. As before, ignoring rotations and reflections, we get 62 decapods. Imagine the convex vertexes on the perimeter of each decapod to be labeled $T$ and the concave vertexes labeled $H$. To continue tiling, these $H$'s and $T$'s must be matched to the heads and tails of the tiles in the usual manner.

When the spokes are arranged the way they are in the infinite cartwheel pattern shown, a decapod called Batman is formed at the center. Batman (shown in dark gray) is the only decapod that can legally be tiled. (No finite region can have more than one legal tiling.) Batman does not, however, force the infinite cartwheel pattern. It merely allows it. Indeed, no finite portion of a legal tiling can force an entire pattern, because the finite portion is contained in *every* tiling.

Note that the infinite cartwheel pattern is bilaterally symmetrical, its axis of symmetry going vertically through Batman. Inflate the pattern and it remains unchanged except for mirror reflection in a line perpen-

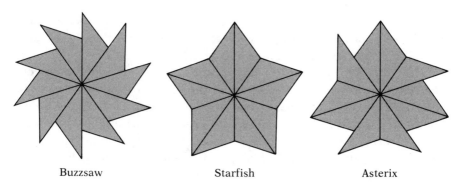

Buzzsaw          Starfish          Asterix

**Figure 12**  *Three decapods*

dicular to the symmetry axis. The five darts in Batman and its two central kites are the only tiles in any Penrose universe that are not inside a region of fivefold symmetry. All other pieces in this pattern or any other one are in infinitely many regions of fivefold symmetry.

The other 61 decapods are produced inside the central cartwheel by the other 61 combinations of worm turns in the spokes. All 61 are "holes" in the following sense. A hole is any finite empty region, surrounded by an infinite tiling, that cannot be legally tiled. You might suppose each decapod is the center of infinitely many tilings, but here Penrose's universes play another joke on us. Surprisingly, 60 decapods force a unique tiling that differs from the one shown only in the composition of the spokes. Only Batman and one other decapod, called Asterix* after a French cartoon character, do not. Like Batman, Asterix allows an infinite cartwheel pattern, but it also allows patterns of other kinds.

Now for a startling conjecture. Conway believes, although he has not completed the proof, that every possible hole, of whatever size or shape, is equivalent to a decapod hole in the following sense. By rearranging tiles around the hole, taking away or adding a finite number of pieces if necessary, you can transform every hole into a decapod. If this is true, any finite number of holes in a pattern can also be reduced to one decapod. We have only to remove enough tiles to join the holes into one big hole, then reduce the big hole until an untileable decapod results.

*Asterix the Gaul is featured in a popular series of French picture books for children. The stories are fantasies taking place at the time of Julius Caesar. Asterix is also an intended pun on "asterisk."

Think of a decapod as being a solid tile. Except for Batman and Asterix, each of the 62 decapods is like an imperfection that solidifies a crystal. It forces a unique infinite cartwheel pattern, spokes and all, that goes on forever. If Conway's conjecture holds, any "foreign piece" (Penrose's term) that forces a unique tiling, no matter how large the piece is, has an outline that transforms into one of 60 decapod holes.

Kites and darts can be changed to other shapes by the same technique described earlier for changing isosceles triangles into spiral-tiling polygons. It is the same technique that Escher employed for transforming polygonal tiles into animal shapes. Figure 13 shows how Penrose changed his darts and kites into chickens that tile only nonperiodically. Note that although the chickens are asymmetrical, it is never necessary to turn any of them over to tile the plane. Alas, Escher died before he could know of Penrose's tiles. How he would have reveled in their possibilities!

By dissecting darts and kites into smaller pieces and putting them together in other ways you can make other pairs of tiles with properties similar to those of darts and kites. Penrose found an unusually simple pair: the two rhombuses in the sample pattern of Figure 14. All edges are the same length. The larger piece has angles of 72 and 108 degrees and

**Figure 13** *Penrose's nonperiodic chickens*

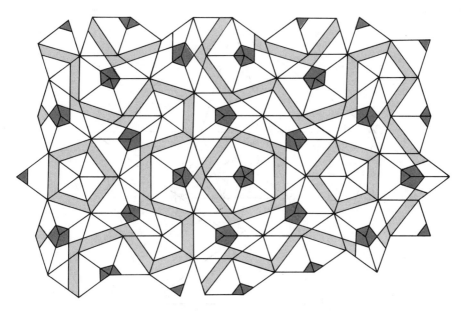

**Figure 14** *A nonperiodic tiling with Roger Penrose's rhombuses*

the smaller one has angles of 36 and 144 degrees. As before, both the areas and the number of pieces needed for each type are in the golden ratio. Tiling patterns inflate and deflate and tile the plane in an uncountable infinity of nonperiodic ways. The nonperiodicity can be forced by bumps and dents or by a coloring such as the one suggested by Penrose and shown in the illustration by the light and dark gray areas.

We see how closely the two sets of tiles are related to each other and to the golden ratio by examining the pentagram in Figure 15. This was the mystic symbol of the ancient Greek Pythagorean brotherhood and the diagram with which Goethe's Faust trapped Mephistopheles. The construction can continue forever, outward and inward, and every line segment is in the golden ratio to the next smaller one. Note how all four Penrose tiles are embedded in the diagram. The kite is *ABCD*, and the dart is *AECB*. The rhombuses, although they are not in the proper relative sizes, are *AECD* and *ABCF*. As Conway likes to put it, the two sets of tiles are based on the same underlying "golden stuff." Any theorem about kites and darts can be translated into a theorem about the Penrose rhombuses or any other pair of Penrose tiles and vice versa. Conway prefers to work with darts and kites, but other mathematicians prefer working with the simpler rhombuses. Robert Ammann has found a

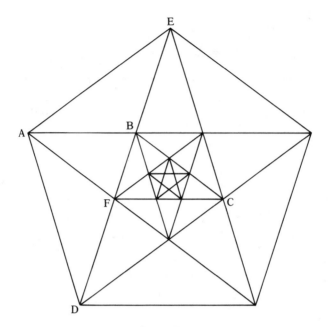

**Figure 15** *The Pythagorean pentagram*

bewildering variety of other sets of nonperiodic tiles. One set, consisting of two convex pentagons and a convex hexagon, forces nonperiodicity without any edge markings. He found several pairs, each a hexagon with five interior angles of 90 degrees and one of 270 degrees. You'll find these sets depicted and their remarkable properties discussed in the book by Branko Grünbaum and G. C. Shephard listed in the next chapter's bibliography.

Are there pairs of tiles not related to the golden ratio that force nonperiodicity? Is there a pair of *similar* tiles that force nonperiodicity? Is there a pair of convex tiles that will force nonperiodicity without edge markings?

Of course, the major unsolved problem is whether there is a *single* shape that will tile the plane only nonperiodically. Most experts think not, but no one is anywhere near proving it. It has not even been shown that if such a tile exists, it must be nonconvex.

# CHAPTER 2
## Penrose Tiling II

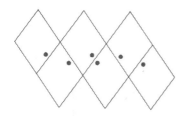

  In the decade since my column on Penrose tiling ran in *Scientific American* (January 1977), Roger Penrose, John Conway, Robert Ammann and others have made enormous strides in exploring nonperiodic tiling. (I will continue here to use the term "nonperiodic," although Branko Grünbaum and G. C. Shephard in their monumental work *Tilings and Patterns* prefer to call a set of tiles "aperiodic" if it tiles only nonperiodically.) The discovery of what are now called Ammann bars or lines and of 3-space analogues of Penrose tiling has led to an amazing development in crystallography, but first let me summarize in this previously unpublished chapter some of the developments that preceded this breakthrough.

  Robert Ammann, a brilliant young mathematician working at low-level computer jobs in Massachusetts, independently discovered Penrose's rhomb tiles in 1976, about eight months before my column on Penrose tiling appeared. In correspondence I informed him of the darts and kites, as well as Penrose's earlier discovery of the rhombs. Ammann soon realized that both pairs of tiles formed patterns that were determined by five families of parallel lines that cross the plane in five different directions, intersecting one another at 360/5 = 72-degree angles. One family of such lines, now called Ammann bars, is shown in Figure 16.

**Figure 16** *A family of Ammann bars displaying* (left to right) *a* SLLSLLS *sequence*

Observe that the lines cross the concave corners of darts that point in the same or opposite direction. This is not strictly accurate, but for our purposes lines ruled in this easy way are adequate. For a precise positioning of the lines, see the Grünbaum/Shephard book. When accurately placed, each line is a trifle outside a dart's concave corner. Inside each regular decagon (ten-sided polygon) on the pattern, the Ammann bars form a perfect pentagram (five-pointed star).

Note that the spacings between bars are of two lengths which we will call $L$ (for long) and $S$ (for short). When the lines are properly drawn, these two lengths are in golden ratio. Moreover, on the infinite plane the number of $L$'s in a family of bars and the number of $S$'s in that same family are in golden ratio. Moving in either direction perpendicular to a family of bars, we can record the sequence of spacings as a sequence of $L$'s and $S$'s. This sequence is nonperiodic, and constitutes a remarkable 1-dimensional analogue of Penrose tiling. The local isomorphism theorem applies. If you select any finite portion of the sequence, you can always find it duplicated not far away. Start anywhere and write down the letters to any finite length, say a billion. If you start at any other spot in the sequence, you are certain to reach this identical billion-letter sequence. Only when the sequence is taken as infinite is it unique.

Conway discovered that this sequence can be obtained from the golden ratio in the following way. Write down in ascending order the multiples of the golden ratio $(1 + \sqrt{5})/2$ and round them down to the nearest integer. The result is the sequence that begins 1, 3, 4, 6, 8, 9, 11, 12, 14, 16, 17, 19, 21, 22, 24, 25, 27, 29, 30, 32, 33, 35, 37, 38, 40, 42, 43, 45, 46, 48, 50. . . . It is sequence 917 in N. J. A. Sloane's *Handbook of Integer Sequences*. If you round down multiples of the square of the golden ratio, you get the sequence 2, 5, 7, 10, 13, 15, 18, 20, 23. . . . The two sequences are called "complementary." Together they display every positive integer once and only once. Successive multiples of any real number $a$, rounded down to the nearest integer, form a sequence called the spectrum of $a$. If $a$ is irrational, the sequence is called a Beatty sequence after Samuel Beatty, a Canadian mathematician who called attention to such sequences in 1926. As we shall see in Chapter 8, the complementary Beatty sequences based on the golden ratio provide the winning strategy for a famous variant of Nim known as Wythoff's game. References on Beatty sequences are given in that chapter's bibliography.

Adjacent numbers in the golden Beatty sequence differ by either 1 or 2. Put down this first row of differences, then change each 1 to 0 and each 2 to 1. You get an endless binary sequence that starts 101101011011010. . . . This is a portion of the sequence of $S$'s and $L$'s in any infinite family of Ammann bars. Conway uses the term "musical sequence" for any finite segment of the golden ratio sequence. Following Penrose, I shall call them Fibonacci sequences.

Such sequences have many curious properties. For example, put a decimal point in front of the Fibonacci sequence given above in binary notation. The result is an irrational binary fraction that is generated by the following continued fraction:

$$\cfrac{1}{1+\cfrac{1}{2+\cfrac{1}{2+\cfrac{1}{2^2+\cfrac{1}{2^3+\cfrac{1}{2^5+\cfrac{1}{2^8+\cfrac{1}{2^{13}+\cfrac{1}{2^{21}+\cfrac{1}{2^{34}+\ \cdots}}}}}}}}}}$$

The exponents of this continued fraction are none other than the Fibonacci numbers. Conway has many unpublished results on the way Penrose tilings are related to Fibonacci numbers, which are in turn related to the growth patterns of plants.

Penrose tilings are, as we have seen, self-similar in the sense that inflating or deflating them produces another tiling. Fibonacci sequences have the same self-similar property. There are many techniques for inflating and deflating them to produce another such sequence, but the simplest is as follows. To deflate, replace each *S* by an *L*, each *LL* by *S*, and drop all single *L*'s. For example, the sequence *LSLLSLSLLSLLSLS* deflates by these rules to *LSLLSLSLL*. To inflate, replace each *L* by *S*, each *S* by *LL*, then add an *L* between each pair of *S*'s.

A Fibonacci sequence cannot contain *SS* or *LLL*. This provides a simple way to tell if a sequence of *S*'s and *L*'s is Fibonacci. Apply the deflation rules until you reach either a sequence that contains an *SS* or an *LLL* (in which case the sequence is not Fibonacci) or a single letter that proves it is. If you inflate or deflate a Penrose tiling, the sequence in each family of Ammann bars also inflates or deflates. The sequence of long and short bow ties in any worm, such as the worms in the ten spokes of the cartwheel pattern, is also a Fibonacci sequence.

Two families of Ammann bars tessellate the plane with nonperiodic parallelograms that form a grid into which the tiles fit. As Grünbaum and Shephard put it, instead of thinking of the tiles as determining Ammann bars, "it is the system of bars which are fundamental and the only function of the tiles is to give a practical realization to them." The bars are something vaguely like the quantum fields that determine the positions and paths of particles. Ammann was the first to perceive, early in 1977, that his grid of bars leads into "forcing theorems"—theorems that tell how a small set of tiles will force the positions of infinite sets of other tiles.

As Ammann expressed it in a letter to me: "Whenever a set of tiles forces two parallel lines to occupy certain positions, it forces an infinite number of nonadjacent parallel lines also to occupy certain positions. Whenever three lines cross at the proper angles, a tile is forced." This property of a finite set of tiles forcing the positions of tiles at arbitrarily long distances belongs also to the Penrose rhombs and to Robinson squares, even though they have no connection with the golden ratio.

Taking off from Ammann's discoveries, Conway went on to develop many remarkable forcing theorems. I will say here only that two Penrose tiles (each can be of either type), suitably placed and arbitrarily far apart, will determine two infinite families (not complete families) of bars.

Intersections of the two families in turn determine the positions of an infinite set of tiles. The king, queen, jack, deuce and star, for example, force an infinite set of tiles in their empires. (The ace and sun do not force any tiles.) The king's empire is unusually dense. You might expect the density of such forced tiles to thin out as you get farther from the center, but this is not the case. The density remains constant for the entire plane.

Ammann's other great discovery, also made in 1976, was a set of two rhombohedra (parallelepipeds with six congruent rhomb faces) which, with suitable face-matching rules, force a nonperiodic tiling of space. Nets for the two solids are shown in Figure 17. If you cut these two nets out of cardboard, fold along the lines and tape the edges, you will obtain the two solids shown at the bottom of the illustration. One can be thought of as a cube that has been squashed along a space diagonal and

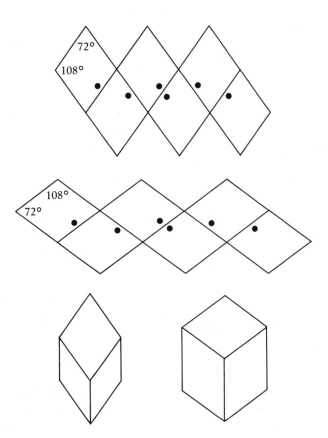

**Figure 17** *Nets for the obtuse and acute golden rhombohedra*

the other as a cube that has been stretched along a space diagonal. All twelve faces are congruent, with their diagonals in golden ratio. The geometer H. S. M. Coxeter, in a note added on page 161 to the thirteenth edition of W. W. Rouse Ball's classic *Mathematical Recreations and Essays* (Dover, 1987), which he edited, calls a rhombohedron of this type a "golden rhombohedron." There are just two kinds, both of which had been studied by Kepler. The acute golden rhombohedron has two opposite corners where three equal acute angles meet. The obtuse golden rhombohedron has two opposite corners where equal obtuse angles meet. Other corners on both solids are mixtures of acute and obtuse angles.

Ammann's two rhombohedra are the two golden types. The faces of the acute solid meet along edges at angles of 72 and 108 degrees. Those on the obtuse solid meet at 36 and 144 degrees. (The four dihedral angles are multiples of $360/10 = 36$ degrees.) The face angles are close to 64 and 116 degrees. Periodic tiling is ruled out by suitably placed holes and projections. Note the spots on the unfolded faces in the illustration. Imagine each solid with a duplicate that has its spots in a pattern that is a mirror image of the other. This forms a set of four solids that force nonperiodicity if you put them together so every spot touches another spot. It is not known if there is a way to avoid this mirror-image marking so that just two solids, suitably marked, will force nonperiodicity. If a plane is passed through the space tiling at a suitable angle, the plane displays a tiling very close to a tiling by Penrose rhombs.

I sent Ammann's results to Penrose. In a letter dated May 4, 1976, Penrose asked me to convey his congratulations to Ammann on two counts: for his independent discovery of the rhomb tiles and for the space tiling by the two golden rhombohedra. He continued:

> It is just possible that these things may have some significance in biology. You will recall that some viruses grow in the shapes of regular dodecahedra and icosahedra. It has always seemed puzzling how they do this. But with Ammann's non-periodic solids as basic units, one would arrive at quasi-periodic 'crystals' involving such seemingly impossible (crystallographically) cleavage directions along dodecahedral or icosahedral planes. Is it possible that the viruses might grow in some such way involving non-periodic basic units—or is the idea too fanciful?

A year after Ammann's discovery of his nonperiodic space tiling, it was rediscovered in Japan by Koji Miyazaki, an architect at Kobe University. He also discovered another way that the two golden rhombohedra can tile space nonperiodically, although the tiling is not forced. Five acute and five obtuse golden rhombohedra will fit together to form a rhombic triacontrahedron. Two such solids, joined by a common obtuse vertex, can be surrounded with 60 more golden rhombohedra (30 of each type) to make a larger rhombic triacontrahedron. This enlargement can be continued to infinity, tiling space in a honeycomb that has a center of icosahedral symmetry.

Penrose's conjectures about crystals, even his terminology, proved to be amazingly prophetic. In the early 1980's a number of scientists and mathematicians began to speculate cautiously about the possibility that the atomic structure of crystals might be based on a nonperiodic lattice. Then in 1984 Dany Schechtman and his colleagues at the National Bureau of Standards made a dramatic announcement. They had found a nonperiodic structure in the electron micrographs of a rapidly cooled aluminum-manganese alloy that some chemists immediately dubbed Schechtmanite. The micrographs displayed a clear fivefold symmetry which strongly suggested a nonperiodic space tiling analogous to Penrose tiling.

Nothing like this had been seen before. It was, as science writer Ivars Peterson put it, as if someone had observed a five-sided snowflake. It had long been a dogma in crystallography that crystals could exhibit rotational symmetry of only 2, 3, 4 and 6 rotations, but never 5, 7 or 8. Another dogma was that solid matter took only two forms: either with atoms in a periodic arrangement or with disordered atoms in such amorphous material as glass.

The ordered lattices of all crystals then known derived from three Platonic solids: the tetrahedron, cube and octahedron. The dodecahedron and icosahedron were ruled out because their fivefold symmetry made periodic tiling impossible. Yet here was a material that seemed to exhibit icosahedral symmetry. Like Penrose tiling, when the material was rotated by 72 degrees, or 1/5 of a circle, it remained essentially the same in an overall statistical way, but without long-range periodicity. It seemed to be a form of matter halfway between glass and ordinary crystals, suggesting that instead of a sharp demarcation between the two forms, there could be a continuum of in-between structures.

Among physicists, chemists and crystallographers the effect of this discovery was explosive. Similar nonperiodic structures were soon being induced in other alloys, and dozens of papers began to appear. It became clear that solid matter could exhibit nonperiodic lattices with any kind of rotational symmetry. Wide varieties of solid tiles in sets of two or more were proposed as models, some forcing nonperiodicity, some merely allowing it. A crystal structure was produced made of layers of sheets with two-dimensional Penrose rhomb tiling. N. G. de Bruijn in the Netherlands developed an algebraic theory of nonperiodic tiling based on what he calls "pentagrids," similar to Ammann bars. In a 1987 paper, he reported a surprising connection between nonperiodic tiling theory and a shuffling theorem known to card magicians as the Gilbreath principle. (On this principle see Chapter 9 of my *New Mathematical Diversions from Scientific American.*)

There is now enormous ferment in the ongoing empirical and theoretical investigations of "quasicrystals," as the new halfway crystals are called. There is also opposition to the view that their lattices are genuinely nonperiodic. The leading opponent is Linus Pauling, who argues that the micrographs should be interpreted as a spurious form of fivefold symmetry known to crystallographers as multiple twinning. "Crystallographers can now cease to worry that the validity of one of the accepted bases of their science has been questioned," Pauling concluded in a 1985 report in *Nature*. Another possibility is that quasicrystals are simply extremely large unit cells of a periodic pattern that will be found when larger samples are made. And there are other possibilities. Proponents of quasicrystals maintain that all these alternative interpretations of the micrographs have been eliminated and that true nonperiodicity is the simplest explanation. It could be that in a few years empirical studies will disconfirm this, and quasicrystals may go the ill-fated way of polywater; but if the nonperiodic interpretation holds, it will be a sensational turning point in crystallography.

Assuming quasicrystals are real, the next few years should see increasingly efficient techniques for producing them. Many questions cry out for answers. What physical forces are involved in the formation of these strange crystals? Penrose has suggested that perhaps nonlocal quantum field effects play a role because without an overall plan it is hard to see how such a crystal could grow in such a way as to preserve its long-range nonperiodic pattern. (In the passage quoted earlier from his 1976 letter, Penrose's speculations about viruses reflected his concern over how a quasicrystal could grow without guidance by nonlocal forces.) What are the elastic and electronic properties of quasicrystals? Will geologists ever find quasicrystals produced by nature?

If quasicrystals are what their defenders think they are, they provide a striking example of how work done in recreational mathematics, purely for fun and aesthetic satisfaction, can turn out to have significant practical applications to the physical world and to technology.

In 1980 I heard Conway lecture on Penrose tiling at Bell Laboratories. Discussing "hole theory," he said he liked to imagine a vast temple with a floor tesselated by Penrose tiles and a circular column exactly in the center. The tiles seem to go under the column. Actually, the column covers a hole that can't be tesselated. Incidentally, on such patterns the Ammann bars get broken out of alignment as they pass through the hole.

A Penrose tiling can, of course, always be colored with four colors so that no two tiles of the same color share a common edge. Can it always be colored with three? It can be shown, Conway said, from the local isomorphism theorem, that if any Penrose tiling is three-colorable, all are, but so far no one has proved that any infinite Penrose tiling is three-colorable.

Conway gave the following simple *reductio ad absurdum* proof (which he credited to Peter Barlow, a British mathematician who died in 1862, best known today for his books of tables) that no tiling pattern can have more than one center of fivefold symmetry. Assume it has more than one. Select the two, *A* and *B*, that are closest together. (See Figure 18.) Rotate the pattern $360/5 = 72$ degrees clockwise around *B*, carrying *A* to *A'* as shown. Return to the original position, and rotate the pattern 72 degrees counterclockwise around *A*, taking *B* to *B'*. Result: Both rotations (if our assumption is true) would leave the pattern unchanged, but now it has two centers of fivefold symmetry, *A'* and *B'*, that are closer together than *A* and *B*. This contradicts our second assumption that *A* and *B* are the closest centers.

There are single tiles (and sets of tiles) that tile the plane periodically in only one way: the regular hexagon and the cross pentomino, for example. All triangles and all parallelograms tile in an uncountable

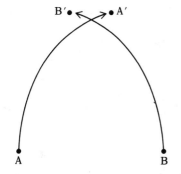

**Figure 18**  *Barlow's proof that no pattern can have two centers of fivefold symmetry*

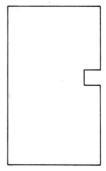

**Figure 19**  *The Conway tile that tiles in zero ways*

infinity of ways. Grünbaum and Shephard conjecture that no tile exists that tiles periodically in a *countable* infinity of ways. They also conjecture that given any positive integer $r$, there are single tiles that tile the plane in just $r$ ways. Such tiles have been found for $r = 1$ through 10. In his lecture Conway exhibited what he calls the "Conway tile" (Figure 19) for $r = 0$. He concluded by saying it was the first lecture he had ever given on Penrose tiling in which he didn't inadvertently say "karts and dites."

## BIBLIOGRAPHY

*Nonperiodic tiling*

"Central Tessellations." Michael Goldberg, in *Scripta Mathematica*, 21, 1955, pp. 253–260.

"Games, Logic and Computers." Hao Wang, in *Scientific American*, November 1965, pp. 98–106.

"The Undecidability of the Domino Problem." Robert Berger in *Memoirs of the American Mathematical Society*, 66, 1966.

"Undecidability and Nonperiodicity for Tilings of the Plane." Raphael M. Robinson, in *Inventiones Mathematicae*, 12, 1971, pp. 177–209.

"The Role of Aesthetics in Pure and Applied Mathematical Research." Roger Penrose, in *Bulletin of the Institute of Mathematics and Its Applications*, 10, 1974, pp. 266–271.

"The Penrose Pieces." Richard K. Guy, in *Bulletin of the London Mathematical Society*, 8, 1976, pp. 9–10.

*On Some Periodical and Non-Periodical Honeycombs.* Koji Miyazaki. A Kobe University monograph, 1977.

"Pentaplexity: A Class of Nonperiodic Tilings of the Plane." Roger Penrose, in *Eureka*, 39, 1978, pp. 16–22. Reprinted in *The Mathematical Intelligencer*, 2, 1979, pp. 32–37, and in *Geometrical Combinatorics*, F. C. Holroyd and R. J. Wilson, eds., Pitman, 1984.

"Algebraic Theory of Penrose's Nonperiodic Tilings of the Plane." N. G. de Bruijn, in *Nederl. Akad. Proc.*, Series A, 84, 1981, pp. 37–66.

"Some Problems on Plane Tilings." Branko Grünbaum and G. C. Shephard, in *The Mathematical Gardner*, edited by David A Klarner. Prindle, Weber & Schmidt, 1981.

"Algebraic Theory of Nonperiodic Tilings of the Plane by Two Simple Building Blocks: a Square and a Rhombus." E. P. M. Beenker, Eindhoven University of Technology, the Netherlands, Department of Mathematics and Computer Science, T. H. Report 82-WSK-04, September 1982.

"Interview with Roger Penrose," in *Omni*, June 1986, pp. 67 ff.

*Tilings and Patterns.* Branko Grünbaum and G. C. Shephard. W. H. Freeman and Company, 1986.

### Quasicrystals

"The Fivefold Way for Crystals." Ivars Peterson, in *Science News*, 127, 1985, pp. 188–189. Reprinted in Peterson's collection *The Mathematical Tourist*, W. H. Freeman and Company, 1988, pp. 200–212.

"Pentagonal and Icosahedral Order in Rapidly Cooled Metals." David R. Nelson and Bertrand I. Halperin, in *Science*, 229, 1985, pp. 235–238.

"Puzzling Crystals Plunge Scientists into Uncertainty." Malcolm W. Browne, in the *New York Times*, Sunday, July 30, 1985, p. C1 ff.

"Apparent Icosahedral Symmetry is Due to Directed Multiple Twinning of Cubic Crystals." Linus Pauling, in *Nature*, 317, 1985, p. 512.

"Pauling's Model Not Universally Accepted." R. Mosseri and J. F. Sadoc, in *Nature*, 319, 1986, p. 104. Reply to Pauling's 1985 paper.

"Quasicrystals: A New Class of Ordered Structure." D. Levine, Ph.D. thesis, University of Pennsylvania, 1986.

"Quasicrystals." David R. Nelson, in *Scientific American*, August 1986, pp. 42–51.

"Quasicrystals." Paul Joseph Steinhardt, in *American Scientist*, November/December, 1986, pp. 586–597. Its bibliography lists more than 40 technical papers.

"A Riffle Shuffle Card Trick and Its Relation to Quasicrystal Theory." N. G. de Bruijn, to be published in *Nieuw Archief voor Wiskunde*.

*The Physics of Quasicrystals*, Paul Steinhardt and Stellan Ostlund, eds. World Scientific, 1987.

"Opening the Door to Forbidden Symmetries." Mort La Brecque, in *Mosaic*, 18, 1987–1988, pp. 2–23.

"Icosahedral Symmetry." A letter from Linus Pauling that continues his attack on quasicrystals, with a reply by Paul Steinhardt, in *Science*, 239, 1988, pp. 963–964.

"Tiling to Infinity." Ivars Peterson, in *Science News*, 134, July 16, 1988, p. 42.

*Introduction to Quasicrystals*. Marko V. Jaric, ed. Academic Press, 1988.

*Introduction to the Mathematics of Quasicrystals*. Marko V. Jaric, ed. Academic Press, 1989.

*Aperiodicity and Order*, Marko V. Jaric and Dennis Gratias, eds. Academic Press, 1989.

"Three-Dimensional Analogs of the Planar Penrose Tilings and Quasicrystals." L. Danzer, in *Discrete Mathematics*, 76, 1989, pp. 1–7.

"Updown Generation of Beatty Sequences." N. G. de Bruijn, in *Proceedings of the Koninklijke Nederlandse Akademie van Wetenschappen*, vol. 92, Series A, December 18, 1989, pp. 385–407.

"Impossible Crystals." Hans C. von Baeyer, in *Discover*, February 1990, pp. 69–78.

"Quasicrystals: Rules of the Game." John Horgan, in *Science*, vol. 247, March 2, 1990, pp. 1020–1022.

"Quasicrystals: The View from Les Houches." Marjorie Senechal and Jean Taylor, in *The Mathematical Intelligencer*, vol. 12, no. 2, 1990, pp. 54–63.

"The Structure of Quasicrystals." Peter V. Stephens and Alan I. Goldman, in *Scientific American*, April 1991, pp. 44–53.

*Quasicrystals.* D. P. Vincenzo and P. J. Steinhardt. World Scientific, 1991.

"Shadows and Symmetries." Ivars Peterson, in *Science News*, vol. 140, December 21 and 26, 1991, pp. 408–410.

*Quasicrystals: A Primer.* C. Janot. Clarendon Press, 1992.

"Penrose Patterns and Quasicrystals." V. Koryepin, in *Quantum*, January/February, 1994, pp. 10–13.

*Quasicrystals and Geometry.* Marjorie Senechal. Cambridge University Press, 1995.

### About Penrose

"Those Computers are Dummies." Michael Lemonick, in *Time*, June 25, 1990, p. 24.

# CHAPTER 3

# Mandelbrot's Fractals

When Zulus cannot smile, they frown,
To keep an arc before the eye.
Describing distances to town,
They say, "As flies the butterfly."

—JOHN UPDIKE, *"Zulus Live in Land
without a Square"*

A fascinating aspect of the history of mathematics is the way that the definitions of names for classes of mathematical objects are continually revised. The process usually goes like this: The objects are given a name, $x$, and defined in a rough way that conforms to intuition and usage. Then someone discovers an exceptional object that meets the definition but clearly is not what everyone has in mind when he calls an object $x$. A new and more precise definition is then proposed that either includes the exceptional object or excludes it. The new definition "works" as long as no new exceptions arise. If they do, the definition has to be revised again, and the process may continue indefinitely.

If the exceptions are strongly counter to intuition, they are sometimes called monsters. The adjective *pathological* is often attached to them. In this chapter we consider the word "curve," describe a few monsters that have forced redefinitions of the term and introduce a frightening new monster captured last year by William Gosper, a brilliant computer scientist now with Symbolics, Inc., in Mountain View, California. Readers of my books have met Gosper before in connection with the cellular-automata game Life. It was Gosper who constructed the "glider gun" that made it possible to "universalize" Life's cellular space. (See the three chapters on Life in my *Wheels, Life, and Other Mathematical Amusements*.)

Ancient Greek mathematicians had several definitions for curves. One was that they are the intersection of two surfaces. The conic-section curves, for instance, are generated when a cone is cut at certain angles by a plane. Another was that they are the locus of a moving point. A circle is traced by a rotating compass leg, an ellipse by a moving stylus that is stretching a closed loop of string around two pins, and so on for other curves generated by more complicated mechanisms.

Seventeenth-century analytic geometry made possible a more precise definition. Familiar curves became the diagrams of algebraic equations. Could a plane curve be defined as the locus of points on the Cartesian plane that satisfy any two-variable equation? No, because the diagrams of some equations emerge as disconnected points or lines, and no one wanted to call such diagrams curves. Calculus suggested a way out. The word "curve" was limited to a graph whose points are a continuous function of an equation.

It seems intuitively obvious that if a curve is a continuous function, it should be possible to differentiate the function or, what amounts to the same thing, to draw a tangent to any point on the curve. In the second half of the nineteenth century, however, mathematicians began to find all kinds of monster curves that had no unique tangent at any point. One of the most disturbing of such monsters was described in 1890 by the Italian mathematician and logician Giuseppe Peano. He showed how a single point, moving continuously over a square, could (in a finite time) pass at least once through every point on the square and its boundary! (Actually any such curve must go through an infinity of points at least three times.) At the limit, the curve becomes a solid square. Peano's curve is a legitimate diagram of a continuous function. Yet nowhere on it can a unique tangent be drawn because at no instant can we specify the direction in which a point is moving.

David Hilbert proposed a simple way to generate a Peano curve with two end points. The first four steps of his recursive procedure should be clear from the pictures in Figure 20. At the limit the curve begins and ends at the square's top corners. The four steps in Figure 21 show how Waclaw Sierpinski generated a closed Peano curve.

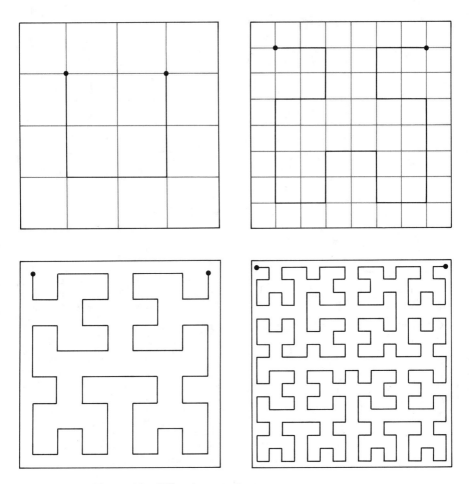

***Figure 20*** *Hilbert's open Peano curve*

In both versions think of the successive graphs as approximations approaching the graph of the limit curve. This limit curve in each version is infinitely long and completely fills the square even though each approximation misses an uncountable infinity of points both of whose coordinates are irrational. (In general the limit of a sequence of approximation curves may go through many points that are not on any of the approximations.) Sierpinski's curve bounds an area 5/12 that of the square. Well, not exactly. The constructions approach this fraction as a limit, but the curve itself, the diagram of the limiting function, abolishes the distinction between inside and outside!

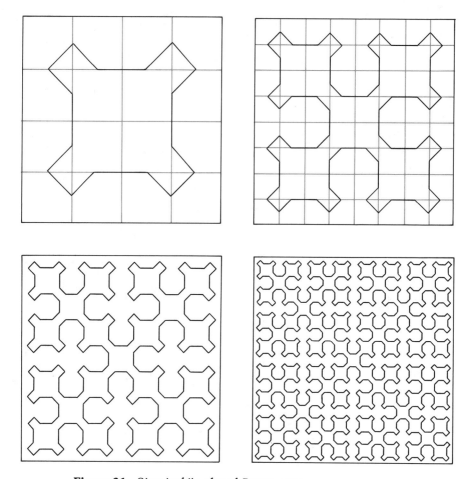

**Figure 21**  *Sierpinski's closed Peano curve*

Peano curves were a profound shock to mathematicians. Their paths seem to be one-dimensional, yet at the limit they occupy a two-dimensional area. Should they be called curves? To make things worse, Peano curves can be drawn just as easily to fill cubes and hypercubes.

Helge von Koch, a Swedish mathematician, proposed in 1904 another delightful monster now called the snowflake curve. We start with an equilateral triangle and apply the simple recursive construction shown in Figure 22 to generate a crinkly curve resembling a snowflake. At the limit it is infinite in length; indeed, the distance is infinite between any two arbitrary points on the curve! The area bounded by the curve is exactly 8/5 that of the initial triangle. Like a Peano curve, its points have no unique tangents, which means that the curve's generating function, although continuous, has no derivative.

If the triangles are constructed inward instead of outward, one gets the anti-snowflake curve. Its perimeter is also infinite, and it bounds an infinity of disconnected regions with a total area equal to 2/5 that of the original triangle. One can start with regular polygons of more than three sides and erect similar polygons on the middle third of each side. A square, with the added squares projecting outward, produces the cross-stitch curve of infinite length that bounds an area twice the original square. (See my *Sixth Book of Mathematical Games from Scientific American*, Chapter 22.) If the added squares go inward, they produce the anti-cross-stitch, an infinite curve that bounds no area. Similar constructions, starting with polygons of more than four sides, produce curves that self-intersect.

A 3-space analogue of the snowflake is constructed by dividing each face of a regular tetrahedron into four equilateral triangles, erecting a smaller tetrahedron on the central triangle and continuing the procedure indefinitely. Will the final result be a finite solid with a surface of infinite area? No, the astonishing answer (Gosper assures me) is that at the limit the surface becomes a perfect cube!

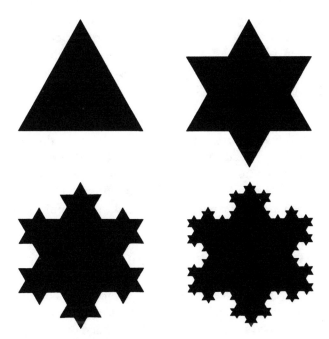

***Figure 22*** *The first four orders of Helge von Koch's snowflake*

We can generalize further by dividing the sides of a regular polygon into more than three parts. For example, divide each side of an equilateral triangle into five parts, erect smaller triangles on the second and fourth sections and repeat to the limit. For an ultimate generalization begin with any closed curve that can be divided into congruent segments, then alter the segments any way you like, provided the alteration is segmented so that the change can be repeated on the smaller segments and carried to the limit. Analogous constructions can be made on the surfaces of solids. Of course, the results may be messy, self-intersecting curves or surfaces of no special interest.

A book could be written about other kinds of pathological planar monsters. The Dutch topologist L. E. J. Brouwer published in 1910 a recursive construction for cutting a region into three subregions in such an insane way that at the limit all three subregions touch at every point (see "Geometry and Intuition," by Hans Hahn; *Scientific American*, April 1954). Brouwer's construction generalizes to divide a region into *n* subregions, all meeting at every point. A more recently discovered family of monsters, the dragon curves, were introduced in *Scientific American*'s Mathematical Games department in 1967 (reprinted in my *Mathematical Magic Show*, Knopf, 1977) and later analyzed by Chandler Davis and Donald Knuth in a 1970 article.

It is now my privilege to present Gosper's new monster, a beautiful Peano curve that he calls the flowsnake. Its construction starts with a pattern of seven regular hexagons (see Figure 23). Eight vertexes are joined as shown by the gray line, made up of seven equal-length segments. The gray line is order 1 of the flowsnake. Order 2, shown in black, is obtained by replacing each gray segment with a similar twisted line of seven segments. Each segment of the black line is $1/\sqrt{7}$ the length of a gray segment; this proportion holds at every stage of the construction.

The recursive procedure is continued to produce flowsnakes of higher orders. Figure 24 shows two computer drawings of flowsnakes of orders 3 and 4. By dividing the plane into black and white, with the bifurcating line passing through the flowsnake's end points, we see how the curve cuts the plane into two regions that twist about in almost, but not quite, the same pattern.

The curve that diagrams the limit of the successive flowsnake functions passes through every point of its region at least once, completely filling the space. The curve is infinite and nondifferentiable. Like the straight line, it is self-similar in the sense that if you enlarge any portion of it, the pattern always looks the same. Snowflake curves have the same property.

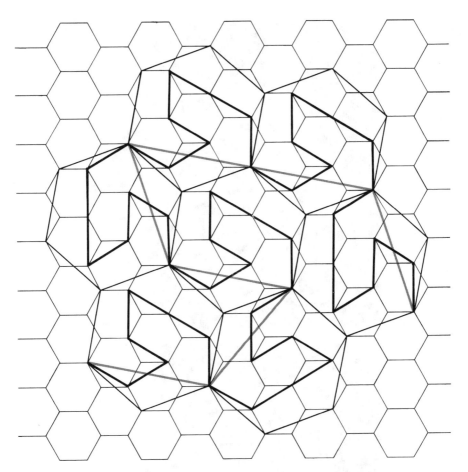

***Figure 23*** *Order 1* (gray) *and order 2* (black) *of William Gosper's "flowsnake"*

In the light of these crazy curves, how do mathematicians currently define a curve? The scene is so crowded with monsters that no single definition covers all the objects to which the word "curve" is commonly applied. The topologist defines a curve as a set of points that are compact and connected and form a 1-dimensional continuum. To make the definition clear, however, a lengthy discourse on point-set topology would be required. The definition catches well-behaved curves that diagram functions with derivatives, but it misses some of the nondifferentiable monsters we have been considering. "Of course we have no physical snowflake curves," Philip Morrison has written. "Nature gives no infinities, not even within molecular collisions. There is a cutoff at the ang-

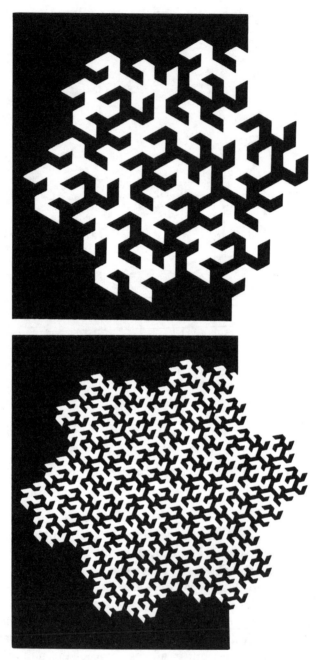

**Figure 24** *Flowsnakes of order 3* (top) *and order 4* (bottom)

strom level. Still, surprises abound." By surprises Morrison means those random natural patterns that have, in a statistical sense, the property of self-similarity as successive enlargements are made. His remarks appear in a review (*Scientific American*, November 1975) of a remarkable French book, *Les Objets Fractals: Forme, Hasard et Dimensions*, by Benoît Mandelbrot.

Mandelbrot is a Polish-born (Warsaw, 1924) French mathematician who is currently an **IBM** Fellow at the Thomas J. Watson Research Center at Yorktown Heights, N.Y. Like Stanislaw Ulam and many other eminent Polish mathematicians, Mandelbrot has had a career involving a marvelous mixture of creative work in both pure and applied mathematics, notably in physics and economics. His teacher, the French mathematician Paul Pierre Lévy, made the first systematic study of statistically self-similar curves, but they were regarded as useless, bizarre curiosities until Mandelbrot recognized them as being a basic tool for analyzing an enormous variety of physical phenomena.

Mandelbrot's book is filled with pictures of just such phenomena. Consider coastlines. Their butterfly-flight irregularity is statistically self-similar. A coastline looks the same from a high altitude as it does from a low one. It is meaningless to speak of a coastline's "length" because it all depends on the precision of measurement. As Morrison puts it, "a coastline on maps at varying scales obeys a power law like the snowflake curve's, from a scale of hundreds of kilometers down to one of perhaps meters, where geography stops and pebbles begin."

The surface of the moon is another example. Remember your surprise on seeing the first closeup photographs of the moon made from a satellite in orbit around it? The moon's pocked surface looked basically as it did in photographs made with telescopes on the earth. Only the crater sizes were different. The same random self-similarity is found on the surface of certain cheeses, in the scattering of stars in the sky, on the bark of trees, in the contours of mountains, in atmospheric turbulence, in auditory noise and in countless other natural patterns. The Brownian motion of suspended particles approximates a statistically self-similar curve that (at the limit) has infinite length and no tangents.

Let us go back to the flowsnake for a close look at its perimeter and at an amazing paradox. The perimeter can be constructed by a recursive procedure much simpler than the one used to get the flowsnake itself. Figure 25 shows how it works. Start with a regular hexagon, then replace each side with a zigzag line (gray) of three equal segments, each $1/\sqrt{7}$ the original side. The result is a nonconvex 18-gon. Since the zigzag line adds the same amount of area as it takes away, the 18-gon obviously has

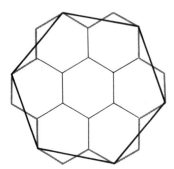

**Figure 25**   *Making the flowsnake's boundary*

the same area as the original hexagon. Repeat the construction on each of the 18 sides to produce a 54-gon, and imagine that the recursive procedure is continued to the limit. At each step the number of sides triples, but the area never changes. At the limit the area filled by the flowsnake is exactly the same as the area of the original hexagon.

The entire region has an astounding property. It can be dissected, as is shown in Figure 26, into seven subregions, each of which is an exact copy of the entire region.

Now for the paradox. What is the ratio of the area of a subregion to the entire region? Clearly it is 1/7, since seven identical subregions make up the whole. But let us approach it from another angle, remembering that the areas of similar figures are proportional to the square of their linear dimensions. The outside perimeter consists of six segments, such as the segment from *A* to *B*, which is half the closed boundary of a subregion. Clearly the boundary of a subregion must be enlarged by a linear factor of 3 to fit the boundary of the entire region. But if this is true, the areas must be in a ratio of $(1/3)^2 = 1/9$. We seem to have proved that the ratio of the areas is both 1/7 and 1/9. As Gosper asked when he first sent the paradox, *Voss ist los?*

The answer lies in the peculiar, counterintuitive character of the pathological boundary. There is no fuzziness about the area of the region it bounds. It is indeed seven times the area of a subregion. Nor is the boundary fuzzy. It *looks* fuzzy, but it is nevertheless a precisely defined infinite set of points. It has, however, a strongly counterintuitive property. By what linear factor must the boundary of one of the seven subregions be enlarged to make it congruent with the outside overall boundary? One would suppose a factor of 3, but the actual factor is $\sqrt{7} = 2.645. \ldots$ It is impossible, of course, to print the boundary because at the limit its complexity is infinite. Only a few steps of its construction can be shown before the ink begins to smear.

A deep question now arises. What "dimension" should be assigned to the flowsnake's boundary? Like the snowflake, it lies in a strange twilight zone between one dimension and two dimensions. In 1919 a German point-set topologist, Felix Hausdorff, resolved the difficulty by giving fractional dimensions to such curves, or what Mandelbrot calls "fractal" dimensions. The term was invented by Mandelbrot about 1975. He based it on the Latin verb *frangere*, which means "to break," and its adjective form, *fractus*. The term suggests the broken, fragmented character of fractals, and also the fractional numbers that provide, as we shall see, a fractal's degree of shagginess. Fractal dimensions should not be confused with Hausdorff spaces — topological structures that mercifully we do not have to go into here.

The familiar Euclidean dimensions 0, 1, 2, 3, 4 . . . are sometimes called topological dimensions because their spaces are topologically

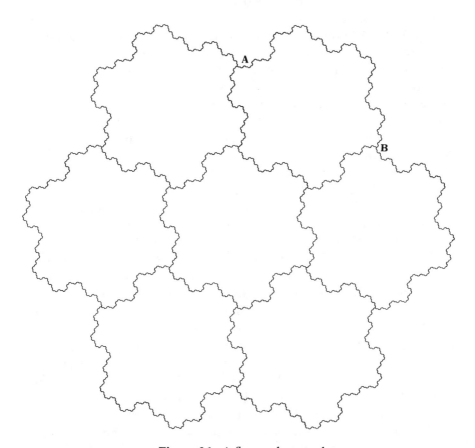

**Figure 26**  *A flowsnake paradox*

distinct; that is, you can't transform one space into another by continuous topological deformation. A point has topological dimension 0. Smooth, well-behaved curves such as straight lines, circles, parabolas and so on have a Euclidean dimension of 1. Surfaces have dimension 2; solids, dimension 3; and hypersolids, higher integers.

To grasp how fractal dimensions are calculated, consider first a straight line segment. If magnified by a factor $x$, the enlarged line can be cut into $y$ copies of the original. The dimension of the line segment is the exponent of $x$ that gives $y$. In this case $x = y$ because doubling the line segment produces two copies of the segment, tripling the line segment produces three copies and so on. We can express the scaling ratio by writing log 2/log 2 = 1.

Magnify a square by doubling its edge. The enlarged square can be cut into four copies of the original. If you triple the edge, it can be cut into nine copies. Generally, if you magnify a plane figure by a linear factor $x$, its area increases by a factor of $x^2$. The dimension of a square, therefore, is log 4/log 2 = 2. If you double a cube's edge, the enlarged cube can be cut into eight copies of the original. Its dimension is log 8/log 2 = 3. And so it goes for hypercubes in higher topological (Euclidean) spaces.

Let us now apply this technique to the snowflake. If you enlarge a portion of it by a linear factor of 3, it produces 4 copies of the original. At each construction step the ragged line is exactly 4/3 times the length of the previous line, although each straight segment is 1/3 the length of the previous segment. It is reasonable, therefore, to assign to the limit curve a Hausdorff dimension $D$, or fractal dimension, that is log 4/log 3 = 1.261859. . . . The boundary of Gosper's flowsnake is constructed by repeatedly replacing a line segment with a zigzag path that is $3/\sqrt{7}$ as long. Its fractal dimension is log 3/log $\sqrt{7}$ = 1.12915. . . .

Calling these numbers "dimensions" is somewhat misleading. They are not Euclidean dimensions. It is best to think of them as measures of complexity, of the "degree of wiggliness," as Mandelbrot once put it. The complexity of Gosper's flowsnake boundary is a trifle lower than the snowflake. Figure 27, from Mandelbrot's 1977 book, is made by replacing lines of four units with lines of eight units to produce a squarish asymmetric snowflake with complexity log 8/log 4 = 1.5. Its fractal dimension, therefore, is a bit higher than the flowsnake boundary. Because each construction step adds the same amount of area as it takes away, the limit curve bounds the same area as the original square.

We can express fractal dimensions with the general formula $D = \log n / \log \sqrt{1/r}$, where $n$ is the number of self-similar parts that result when the original is magnified by factor $r$ and $D$ is the fractal dimension. Fractal

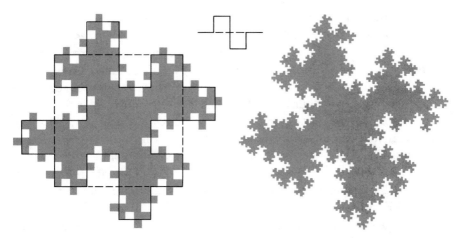

**Figure 27** *The first three orders of Benoît Mandelbrot's square snowflake*

curves that are 1-dimensional in the Euclidean sense can have a fractal number that ranges from 1 to 2 if the curve is on the plane, but can go to higher numbers if the curve twists through higher Euclidean spaces. If a fractal curve passes through every point on the plane within its boundary, as does the flowsnake (the snake itself, not its boundary) and other Peano curves, then it achieves at the limit both a Euclidean dimension and a fractal dimension of 2. If it passes through every point in a solid, it achieves at the limit Euclidean and fractal dimensions of 3. Similarly, a fractal surface of Euclidean dimension 2 can have a fractal dimension that ranges from 2 to 3 as long as it is confined to 3-space, but can go higher if it twists through higher Euclidean spaces, and so on for fractal structures of higher topological dimensions.

Can fractals have complexity numbers less than 1? The answer is yes, although the structures are not topologically equivalent to line segments. For example, remove the center third of a line segment, then remove the central third of each of the two remaining segments and continue this process of third removing to infinity. The result is what Mandelbrot likes to call "Cantor dust." In the literature they are called Cantor discontinua or Cantor sets, after Georg Cantor, who studied them. When you apply Mandelbrot's formula for fractals, you get a number between 0 and 1. In this case, the number is $D = \log 2/\log 3 = 0.6309. \ldots$ Other "cutout" procedures give other numbers. Mandelbrot calls them "subfractals" to distinguish them from fractals with numbers higher than their Euclidean dimension.

In the abstract world of pure mathematics, fractal structures such as we have considered are called "ordered fractals." In the real world there are, of course, no ordered fractals. Structures such as coastlines, trees, rivers, clouds, blood vessels, lightning bolts, paths of particles in Brownian motion and thousands of other fractal-like phenomena are imperfect models that are (within certain upper and lower bounds) fractals in a statistical sense. They are self-similar in that they preserve a statistical similarity that is independent of scaling. One must average the fractal numbers at different scalings, and of course to do this one must make empirical studies. Such fractals are called random fractals or statistical fractals. Coastlines, for instance, have fractal dimensions that change from coast to coast. Investigations show that they fall within a range of 1.15 to 1.25, the second number measuring the complexity of the west coast of England.

The surfaces of mountains are splendid models of random surface fractals. Mandelbrot and his associates, notably Richard F. Voss at IBM, have for the past few years been writing computer programs that generate artificial mountain ranges, clouds and imaginary planets with artificial oceans and continents. Plate 2 shows one of Voss's most striking computer displays: an earthlike planet that never was, as viewed from an equally artificial moon. Artificial clouds have also been produced by computer programs based on fractal formulas. Trees have proved to be more difficult to mimic, but Michael Barnsley and his colleagues at the Georgia Institute of Technology have found ways to mimic leaves and ferns (see Ivars Peterson's 1987 article). These computer graphic displays have led to the generation of strange artificial landscapes for imaginary planets in science fiction films, starting with *Star Trek II: The Wrath of Khan*.

Cantor sets can be constructed in any Euclidean space to make spongelike structures of "dust" with fractal numbers less than the number of the space. Mandelbrot has produced random fractals of Cantor dust in 3-space that model to a surprising degree the apparent distribution of stars in the universe. The hierarchy of star clusters, superclusters and super-superclusters suggests that perhaps the entire universe is, within limits, close to the structure of a random fractal.

## ADDENDUM

Although this chapter has been considerably expanded and updated since it first appeared in *Scientific American* in 1976, some additional updating is called for.

In 1967, when Mandelbrot first wrote about fractals in his classic paper "How Long is the Coast of Britain?," he surely could not have anticipated the speed with which his work would trigger revolutions in both mathematics and physics. Not only have fractals become one of the most energetic research areas of topology, with high-speed computers serving as essential tools, they have also become fundamental aspects of a revolutionary new field of physics called chaos theory. The topic is so vast and developing so rapidly that I can do no more here than refer the reader to the first nontechnical book on the topic, James Gleick's splendid *Chaos: Making a New Science* (Viking, 1987).

Gleick's survey will introduce you to the most important fractals involved in what are called the "strange attractors" of chaos, to such beautiful fractals (on the complex plane) as Julia sets and to the incredible Mandelbrot set (M-set) that Mandelbrot found in 1980. It has been called the most mysterious object in geometry. As research continues, the definition of fractal has broadened to a point at which Mandelbrot has proposed that a final definition be postponed until things settle down. It is no longer necessary, for example, that ordered fractals preserve self-similarity at all scalings. All sorts of "nonlinear fractals" have been constructed, such as the affine fractals that show affine distortions at successive magnifications. The M-set, a sort of dictionary for all Julia sets (I won't try to explain what *that* means!), is not self-similar, except in a topological way, although it contains an infinite number of copies of itself. Every new level of enlargement reveals unpredictable surprises. For more than a year, as successive magnifications were made, it was not even known if the set is connected. Each enlargement disclosed isolated "islands" and particles of dust. Further enlargements would join these disconnected portions to the mainland, but new islands and dust would appear. It was finally proved in 1982 that the set is indeed connected at the limit, but it may be decades before its major properties are uncovered.

In an interview in *Omni* (June 1986) the British mathematical physicist Roger Penrose invoked the Mandelbrot set to support his Platonic approach to mathematics:

> Have you ever seen those pictures produced by computers, the object known as the Mandelbrot set? It's as if you are traveling to some distant world. You turn on your sensing device and see this incredibly complicated configuration, with all sorts of structure to it, and you try to figure out what it is. You might think it is some extraordinary landscape or perhaps some kind of creature with lots of little babies all over the place, babies that are almost but not quite the same as the creature

itself. Very elaborate and impressive! Yet just from seeing the equations, nobody had the remotest conception that they would produce patterns of this nature. Now these landscapes aren't conjured up out of someone's imagination; everyone sees the same pattern. You're exploring something with a computer, but it's not dissimilar from exploring something with experimental apparatus.

If you know your way around the complex plane and are interested in exploring by computer the amazing jungles of the M-set, you should subscribe to *Amygdala*. This is a monthly newsletter that reports new discoveries about the M-set, more efficient computer techniques for investigating it and anything else related to the M-set that strikes the editor's fancy. It has even introduced a new subgenre of science fiction called M-set SF, centered around such notions as that the M-set is a living entity inhabiting hyperspacetime and possessing paranormal powers. Rollo Silver is the newsletter's founder and editor. He describes himself as "an ontological engineer who lives and works in the mountains of Northern New Mexico. Deprived of the company of his peers and half-crazed by isolation, he started *Amygdala* in self-defense in 1986."

You can get a flyer containing samples of various issues and information on how to subscribe to the newsletter and to a color-slide supplement by writing to *Amygdala*, Box 219, San Cristobel, NM 87564. *Amygdala*, by the way, is Latin for almond. The title honors Mandelbrot, whose name in German and Yiddish means "almond bread."

Mandelbrot's *The Fractal Geometry of Nature* is surely one of the most beautiful, witty and exciting books about mathematics ever published. Its text and breathtaking graphics catch such marvelous monsters as the Devil's staircase, Minkowski sausages, Gosper's fudgeflake, Bernoulli clusters, Sierpinski carpets, Menger sponges, Fatou dust, squigs, dragons and all kinds of curds and cheeses.

Since 1987 Mandelbrot has been a professor of mathematics at Yale University. In 1980 he was given the F. Bernard Medal for Meritorious Service to Science by Columbia University, a prestigious award recommended every five years by the National Academy of Sciences. Since then he has received six other distinguished service awards and six honorary doctorates. There are sure to be more honors to come. He has been called the most versatile mathematician since John von Neumann and Norbert Wiener.

## BIBLIOGRAPHY

### Books and articles by Benoit Mandelbrot

"How Long Is the Coast of Britain?" *Science*, 156, 1967, pp. 636–638.
*Les Objets Fractals: Forme, Hasard et Dimensions*. Paris: Flammarion, 1975.
*Fractals: Form, Chance, and Dimension*. W. H. Freeman and Company, 1977. Translation of previous item.
"Getting Snowflakes into Shape." *New Scientist*, June 22, 1978, pp. 808–810.
*Fractals and the Geometry of Nature*. The Encyclopedia Brittanica's *1981 Yearbook of Science and the Future*.
*The Fractal Geometry of Nature*. W. H. Freeman and Company, 1982. Revised and much expanded version of the 1977 Freeman book.

### Books about fractals

*The Geometry of Fractal Sets*. Kenneth Falconer. Cambridge University Press, 1985.
*The Beauty of Fractals*. Heinz-Otto Peitgen and Peter Richter. Springer-Verlag, 1986. Contains 40 pages of full-color pictures.
*The Science of Fractal Images*. Heinz-Otto Peitgen, ed. Springer-Verlag, 1987.
*Fractals*. Jens Feder. Plenum, 1988.
*Fractals in the Natural Sciences*. R. Ball and M. Fleischman, eds. Princeton University Press, 1988.
*Fractals Everywhere*. Michael Barnsley. Academic Press, 1988.
*The Geometry of Fractal Sets*. Kenneth Falconer. Cambridge University Press, 1989.
*Fractals for the Classroom* (four volumes). Heinz-Otto Peitgen, et al. Springer-Verlag, 1989.
*Fractal Geometry*. Kenneth Falconer. Wiley, 1990.
*Fractals in Physics*. J. Feder and A. Aharony, eds. Elsevier, 1990.
*Computers and the Imagination*. Clifford Pickover. St. Martin's Press, 1991.
*Fractals*. Hans Lauwerier. Princeton University Press, 1991.
*The Pattern Book*. Clifford Pickover. World Scientific, 1995.
*Keys to Infinity*. Clifford Pickover. Wiley, 1995.

### Nontechnical articles and interviews

"Number Representations and Dragon Curves, Parts 1 and 2." Chandler Davis and Donald Knuth, in *Journal of Recreational Mathematics*, 3, 1970, pp. 68–81, 133–149.
"Fractals: A World of Nonintegral Dimensions." Lynn Arthur Steen, in *Science News*, 112, 1977, pp. 122–123.
"Fractal Music." Martin Gardner, in *Scientific American*, April 1978, pp. 16–32.
"A Place in the Sun for Fractals." Dietrick Thomsen, in *Science News*, 121, 1981, pp. 28–30.
"A New Geometry of Nature." Bruce Schechter, in *Discover*, June 1982, pp. 66–67.
"The Fractal Cosmos." Kathleen Stein, in *Omni*, February 1983, pp. 63 ff.
"Geometric Forms Known as Fractals Find Sense in Chaos." Jeanne McDermott, in *Smithsonian*, 14, December 1983, pp. 110–175.

"The Fractal Geometry of Mandelbrot." Anthony Barcellos, in *The College Mathematics Journal*, 15, 1984, pp. 98–177.

"Ants in Labyrinths and Other Fractal Excursions." Ivars Peterson, in *Science News*, 125, 1984, pp. 42–43.

"Interview with Benoit B. Mandelbrot." Monte Davis, in *Omni*, February 1984, pp. 65–66, 102–107.

"Benoit Mandelbrot," An interview by Anthony Barcellos, in *Mathematical People*, edited by Donald Albers and G. L. Alexanderson (Birkhauser, 1985).

"Exploring the Mandelbrot Set." A. K. Dewdney, in *Scientific American*, August 1985, pp. 16–24.

"The Man Who Reshaped Geometry." James Gleick, in the *New York Times Magazine*, December 8, 1985, pp. 64 ff.

"Of Fractal Mountains, Graftal Plants and Other Computer Graphics at Pixar." A. K. Dewdney, in *Scientific American*, 255, December 1986, pp. 14–20.

"Fractal Growth." Leonard Sander, in *Scientific American*, January 1987, pp. 94–100.

"Packing It In; Fractals Play an Important Role in Image Compression." Ivars Peterson, in *Science News*, 131, 1987, pp. 283–285.

"Fractal Fairy Tales." Bruce Schechter, in *Omni*, October 1987, pp. 87–91.

"Physiology in Fractal Dimensions." Bruce West and Ary Goldberger, in *American Scientist*, 75, 1987, pp. 354–365.

*The Mathematical Tourist: Snapshots of Modern Mathematics*, Chapters 5 and 6. Ivars Peterson. W. H. Freeman and Company, 1988.

"The Language of Fractals." Hartmut Jürgens, et al., in *Scientific American*, August 1990, pp. 60–68.

"A Ride on Sierpinski's Carpet." I. M. Sokolov, in *Quantum*, May/June 1992, pp. 7–11.

### Books on Fractals and Chaos Theory

*Chaos, Fractals, and Dynamics*. P. Fisher and W. Smith, eds. M. Decker, 1985.

*Chaos*. James Gleick. Viking, 1987.

*Chaos and Fractals*. Robert Devaney and Linda Keen. American Mathematical Society, 1989.

*Turbulent Mirror*. John Briggs and David Peat. Harper & Row, 1989.

*Computers, Pattern, Chaos, and Beauty*. Clifford Pickover. St. Martin's Press, 1990.

*Fractals, Chaos, Power Laws*. Manfred Schroeder. W. H. Freeman, 1991.

*Chaos and Wonderland*. Clifford Pickover. St. Martin's Press, 1994.

*Nature's Chaos*. Eliot Porter and James Gleick. Viking, 1990.

*Chaos and Fractals*. Heinz-Otto Peitgen, et al. Springer-Verlag, 1992.

*The Collapse of Chaos*. Jack Cohen and Ian Stewart. Viking, 1994.

# CHAPTER 4

# Conway's Surreal Numbers

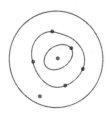

Some said "John, print it";
   others said, "Not so."
Some said "It might do good";
   others said, "No."

—JOHN BUNYAN, *Apology for His Book*

John Horton Conway, the almost legendary mathematician of the University of Cambridge (now at Princeton University), quotes the above lines at the end of the preface to his book, *On Numbers and Games* (Academic Press, 1976), or *ONAG*, as he and his friends call it. It is hard to imagine a mathematician who would say not so or no. The book is vintage Conway: profound, pathbreaking, disturbing, original, dazzling, witty and splattered with outrageous Carrollian wordplay. Mathematicians, from logicians and set theorists to the humblest amateurs, will be kept busy for decades rediscovering what Conway has left out or forgotten and exploring the strange new territories opened by his work.

The sketch of Conway reproduced here could be titled "John 'Horned' (Horton) Conway." The infinitely regressing, interlocking horns form, at the limit, what topologists call a "wild" structure; this one is termed the Alexander horned sphere. Although it is equivalent to the

*John "Horned" (Horton) Conway, as sketched by a colleague on computer-printout paper*

simply connected surface of a ball, it bounds a region that is not simply connected. A loop of elastic cord circling the base of a horn cannot be removed from the structure even in an infinity of steps. (A four-horned mechanical puzzle sold as Loony Loop has a nylon loop that *can* be removed.)

Conway is the inventor of the computer game Life, which I had the honor of introducing in my *Scientific American* columns in 1971 (see the three chapters on Life in my *Wheels, Life, and Other Mathematical Amusements*, W. H. Freeman and Company, 1983). By carefully choosing a few ridiculously simple transition rules Conway created a cellular automaton structure of extraordinary depth and variety. Now he has done it again. By invoking the simplest possible distinction—a binary division between two sets—and adding a few simple rules, definitions and conventions, he has constructed a rich field of numbers and an equally rich associated structure of two-person games.

The story of how Conway's numbers are created on successive "days," starting with the zeroth day, is told in Donald E. Knuth's novelette *Surreal Numbers* (Addison-Wesley, 1974). Since I discussed Knuth's book in the chapter "Nothing" in my *Mathematical Magic Show* (Knopf, 1977), I shall say no more about it here except to remind readers that the construction of the numbers is based on one rule: If we are given a left set $L$ and a right set $R$ and no member of $L$ is equal to or greater than any member of $R$, then there is a number $\{L|R\}$ that is the "simplest number" (in a sense defined by Conway) in between.

By starting with literally nothing at all (the empty set) on the left and the right, $\{\ |\ \}$, one obtains a definition of zero. Everything else follows by the technique of plugging newly created numbers back into the left-right arrangement. The expression $\{0|0\}$ is not a number, but $\{0|\ \}$, with the null set on the right, defines 1, $\{\ |0\}$ defines $-1$ and so on.

Proceeding inductively, Conway is able to define all integers, all integral fractions, all irrationals, all of Georg Cantor's transfinite numbers, a set of infinitesimals (they are the reciprocals of Cantor's numbers, not the infinitesimals of nonstandard analysis) and infinite classes of weird numbers never before seen by man, such as

$$\sqrt[3]{(\omega + 1)} - \frac{\pi}{\omega}$$

where $\omega$ is omega, Cantor's first infinite ordinal.

Conway's games are constructed in a similar but more general way. The fundamental rule is: If $L$ and $R$ are any two sets of games, there is a game $\{L|R\}$. Some games correspond to numbers and some do not, but all of them (like the numbers) rest on nothing. "We remind the reader again," Conway writes, "that since ultimately we are reduced to questions about members of the empty set, no one of our inductions will require a 'basis.'"

In a "game" in Conway's system two players, Left and Right, alternate moves. (Left and Right designate players, such as Black and White or Arthur and Bertha, not who goes first or second.) Every game begins with a first position, or state. At this state and at each subsequent state a player has a choice of "options," or moves. Each choice completely determines the next state. In standard play the first person unable to make a legal move loses. This is a reasonable convention, Conway writes, "since we normally consider ourselves as losing when we cannot find any good move, we should obviously lose when we cannot find any move at all!" In "misère" play, which is usually much more difficult to analyze, the person who cannot move is the winner. Every game can be diagrammed as a rooted tree, its branches signifying each player's options at each successive state. On Conway's trees Left's options go up and to the left and Right's go up and to the right.

Games may be "impartial," as in Nim, which means that any legal move can be made by the player whose turn it is to move. If a game is not impartial, as in chess (where each player must move only his own pieces), Conway now calls it a partizan game. His net thus catches both an enormous variety of familiar games, from Nim to chess, and an infinity of games never before imagined. Although his theory applies to games with an infinity of states or to games with an infinity of options or to both, he is concerned mainly with games that end after a finite number of moves. "Left and Right," he explains, "are both busy men, with heavy political responsibilities."

Conway illustrates the lower levels of his theory with positions taken from a partizan domino-placing game that I discussed briefly in *Knotted Doughnuts and Other Mathematical Entertainments* (W. H. Freeman and Company, 1986), Chapter 19, as the game of Crosscram. (Conway calls it Domineering.) The board is a rectangular checkerboard of arbitrary size and shape. Players alternately place a domino to cover two adjacent squares, but Left must place his pieces vertically and Right must place his horizontally. The first player who is unable to move loses.

An isolated empty square

allows no move by either player. "No move allowed" corresponds to the empty set, so that in Conway's notation this simplest of all games is assigned the value { | } = 0, the simplest of all numbers. Conway calls it Endgame. Its tree diagram, shown at the right of the square, is merely the root node with no branches. Because neither side can move, the second player, regardless of whether he is Left or Right, is the winner. "I courteously offer you the first move in this game," writes Conway. Since you cannot move, he wins.

A vertical strip of two (or three) cells

offers no move to Right but allows one move for Left. Left's move leads to a position of value 0, so that the value of this region is {0| } = 1. It is the simplest of all positive games, and it corresponds to the simplest positive number. Positive games are wins for Left regardless of who starts. The region's tree diagram is shown at the right above.

A horizontal strip of two (or three) cells

allows one move for Right but no move for Left. The value of the region is { |0} = −1. It is the simplest of all negative games and corresponds to the simplest negative number. Negative games are wins for Right regardless of who starts.

A vertical strip of four (or five) cells

has a value of 2. Right has no moves. Left can, if he likes, take the two middle cells in order to leave a zero position, but his "best" move is to take two end cells, because that leaves him an additional move. If this region is the entire board, then of course either play wins, but if it is an isolated region in a larger board or in one of many boards in a "compound game," it may be important to make the move that maximizes the number of additional moves it leaves for the player. For this reason the tree shows only Left's best line of play. The value of the game is $\{1,0| \;\} = \{1| \;\} = 2$. A horizontal strip of four cells has a value of $-2$. If only one player can move in a region and he can fit $n$ of his dominoes into it but no more, then clearly the region has the value $+n$ if the player is Left and $-n$ if the player is Right.

Things get more interesting if both players can move in a region, because then one player may have ways of blocking his opponent. Consider the following region:

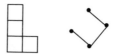

Left can place a domino that blocks any move by Right, thus leaving a zero position and winning. Right cannot similarly block Left because Right's only move leaves a position of value 1. In Conway's notation the value of this position is $\{0, - 1|1\} = \{0|1\}$, an expression that defines $1/2$. The position therefore counts as half a move in favor of Left. By turning the $L$ region on its side one finds that the position is $\{ - 1|0,1\} = \{ - 1|0\} = -1/2$, or half a move in favor of Right.

More complicated fractions arise in Conway's theory. For example,

has a value of $\{1/2|1\} = 3/4$ of a move for Left, since $3/4$ is the simplest number between $1/2$ and 1, the values of the best options for Left and Right. In a game called partizan Hackenbush (the impartial form of this game was explained in my *Wheels, Life, and Other Mathematical Amusements* [cited earlier], and is more fully treated in Conway's book), Conway gives an example of a position in which Left is exactly $5/64$ move ahead!

The values of some game positions are not numbers at all. The simplest example is illustrated in Crosscram by this region:

Both Left and Right have opening moves only, so that the first person to play wins regardless of whether he is Left or Right. Since each player can reduce the value to 0, the value of the position is {0|0}. This is not a number. Conway symbolizes it with * and calls it "star." Another example would be a Nim heap containing a single counter. It is the simplest "fuzzy" game. Fuzzy values correspond to positions in which either player can win if he moves first.

The value of a compound game is simply the sum of the values of its component games. This statement applies also to the value of a position in a game in progress that has been divided by play into a set of subgames. For example, Figure 28 shows a position in a game of Crosscram played on a standard chessboard. The values of the isolated regions are indicated. The position seems to be well balanced, but the regions have a sum of 1¼, which means that Left is one and one-fourth of a move ahead and therefore can win regardless of who moves next. This outcome

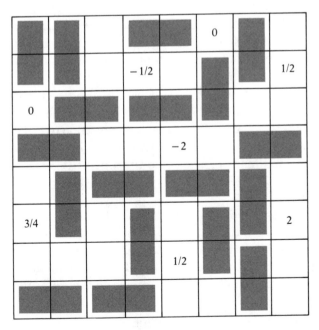

**Figure 28** *A Crosscram position that will result in a win for Left*

would be tedious to decide by drawing a complete tree, but Conway's theory gives it quickly and automatically.

A game is not considered "solved" until its outcome (assuming that both players make their best moves) is known (that is, whether the game's value is zero, positive, negative or fuzzy) and a successful strategy is found for the player who has the win. This stricture applies only to games that must end, but such games may offer infinite options, as in the game Conway calls "My Dad Has More Money than Yours." Players alternately name a sum of money for just two moves and the highest sum wins. Although the tree, Conway admits, is complicated, the outcome is clearly a second-player win.

Are these not trivial beginnings? Yes, but they provide a secure foundation on which Conway, by plugging newly created games back into his left-right scheme, carefully builds a vast and fantastic edifice. I shall not proceed further with it here; instead I shall describe a few unusual games that Conway analyzes in the light of his theory. In all these games we assume standard play in that the first person who is unable to move loses.

1. Col (named for its inventor, Colin Vout). A map is drawn on brown paper. $L$ has black paint, $R$ has white. They alternately color a region with the proviso that regions sharing a border segment not be the same color. It is useful to regard all regions bordering a white one as being tinted white and all regions bordering a black one as being tinted black. A region acquiring both tints drops out of the map as being an unpaintable one.

Conway analyzes Col on the map's dual graph (see Figure 29), defining what he calls "explosive nodes" and marking them with lightning bolts. Of course, the game can be played on white paper with pencils of any two colors. Vout has reported that in the set of all topologically distinct connected maps of one region through five regions 9 are first-player wins and 21 are second-player wins. The game in general is unsolved.

2. Snort (after Simon Norton). This is the same as Col except that neighboring regions must be the same color. It too is unsolved. Conway suspects it has a richer theory than Col. His most valuable tip: If you can color a region adjacent to every region of your opponent's color, do so. In addition Conway has discovered some basic theorems that he reports under the heading "A Short Snort Dictionary."

3. Silver Dollar Game without the Dollar. The board is a horizontal strip of cells extending any length to the right (see Figure 30A). Pennies are arbitrarily placed on certain cells, one to a cell, to provide the initial

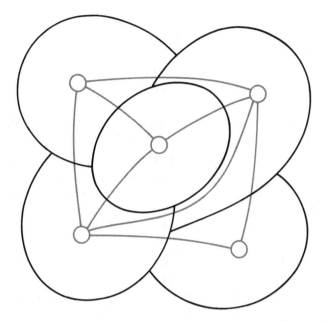

**Figure 29** *A five-region map, with its dual graph* (gray), *for playing Col or Snort*

position. Players alternately move a coin to the left to any empty cell. No jumps are allowed. Eventually all the coins jam at the left, and the person who cannot move loses.

The game is simply Nim in one of its endless disguises. (I assume the reader is familiar with Nim and knows how to determine the winning

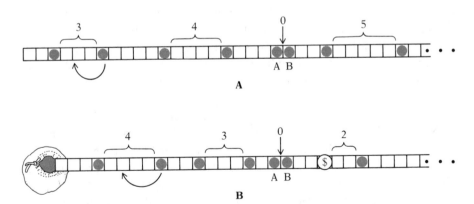

**Figure 30** *(A) The silver-coin game without the silver dollar* *(B) The same game with the dollar*

strategy. If not, he can consult any number of books, including Conway's book or my *Scientific American Book of Mathematical Puzzles & Diversions* [Simon and Schuster, 1959]). Corresponding to the Nim heaps (or rows) of counters are the vacant cells between pennies, starting with the vacancy at the extreme right and including only alternate vacancies. In the illustration the Nim heaps are indicated by brackets and one arrow. The heaps are 3, 4, 0 and 5, so that the game is equivalent to playing Nim with rows of three, four and five counters.

Rational play is exactly as in Nim: Move to reduce the Nim sum of the heaps to 0, a game with a second-player win. The one trivial difference is that here a heap can increase in size. If, however, you have the win and your opponent makes such a move, you immediately restore the heap to its previous size by moving the coin that is just to the right of the heap.

If in the illustrated position it is your move, you are sure to win if you make the move indicated by the curved arrow. If your opponent responds by moving counter *A* two cells to the left, the move raises the empty heap to 2. Your response is to move *B* two cells to the left, thus returning the heap to 0.

4. *Silver Dollar Game with the Dollar.* This is the same as the preceding game except that one of the coins (any one) is a silver dollar and the cell farthest to the left is a money bag (see Figure 30B). A coin farthest to the left can move into the bag. When the dollar is bagged, the game ends and the next player wins by taking the bag.

This game too is Nim disguised. Count the bag as being empty if the coin at its right is a penny and full if it is the dollar, and play Nim as before. If you have the win, your opponent will be forced to drop the dollar in the bag. If the winner is deemed to be the player who bags the dollar, count the bag as being full if the coin at its right is the coin just to the left of the dollar and count it as being empty otherwise. The position shown corresponds to Nim with heaps of 4, 3, 0 and 2. The first player wins in both versions only if he makes the move indicated by the curved arrow.

5. *Rims.* The initial position of this pleasant way of playing a variant of Nim consists of two or more groups of spots. The move is to draw a simple closed loop through any positive number of spots in one group. The loop must not cross itself or cross or touch any other loop. A game is shown in Figure 31.

Conway shows that Rims is the same as Nim played with the added rule that you are allowed, if you like, to take from the center of a row, leaving two new rows instead of one or more rows. Although the number

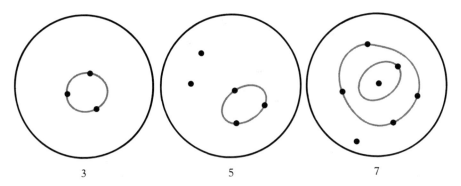

**Figure 31**  *A position in Rims*

of heaps can increase, the winning strategy is standard Nim strategy. If each loop is confined to one or two spots, the game is equivalent to the familiar game of Kayles. (See Chapter 16 of my *Mathematical Carnival*, Knopf, 1975.) Conway calls it Rayles.

6. Prim and Dim. Let me introduce these games by explaining Prime Nim, a simpler game not discussed by Conway. It was first analyzed some 20 years ago by Claude E. Shannon. Prime Nim is played the same way that Nim is except players must diminish heaps only by prime numbers, including 1 as a prime. "The game is actually a bit of a mathematical hoax," wrote Shannon (in a private communication), "since at first sight it seems to involve deep additive properties of the prime numbers. Actually the only property involved is that all multiples of the first nonprime are nonprime."

The first nonprime is 4. The strategy therefore is merely to regard each heap as being equal to the remainder when its number is divided by 4 and then to play standard Nim strategy with these modulo-4 numbers. If 1 is not counted as being prime, the strategy is less simple in standard play and is so complicated in misère play that I do not believe it has ever been solved.

Prim, suggested by Allan Tritter, requires that players take from each heap a number prime to the heap's number. In other words, the two numbers must not be equal or have a common divisor other than 1. Dim requires that a player remove a *divisor* of n (including 1 and n as divisors) from a heap of size n. Conway gives solutions to both games as well as variants in which taking 1 from a heap of 1 is allowed in Prim and taking n from a heap of n is not allowed in Dim.

7. Cutcake. This is a new partizan game invented by Conway. It is played with a set of rectangular cakes, each scored into unit squares like a waffle. Left's move is to break a piece of cake into two parts along any horizontal lattice line, Right's is to break a piece along any vertical line. The game has a surprisingly simple theory.

Figure 32 shows a 4 × 7 cake. In Conway's notation its value is 0, which means it is a second-player win regardless of who goes first. It looks as if the vertical breaker, who has twice as many opening moves as his opponent, would have the advantage, but he does not if he goes first. Assume that the vertical breaker goes first and breaks along the line indicated by the arrow. What is the second player's winning response?

I have given only a few examples of Conway's exotic nomenclature. Games can be short, small, all small, tame, restive, restless, divine, extraverted and introverted. There are ups, downs, remote stars, semistars and superstars. There are atomic weights and sets with such names as On, No, Ug and Oz. Conway has a temperature theory, with thermographs on which hot positions are cooled by pouring cold water on them. He has a Mach principle for the small world: the atomic weight of a short all-small game is at least 1 if and only if the game exceeds the remote stars!

Conway's theorem 99 conveys some notion of the book's whimsical flavor. It tells us (I paraphrase to remove a slight error that Conway discovered too late to correct) that any short all-small game of atomic

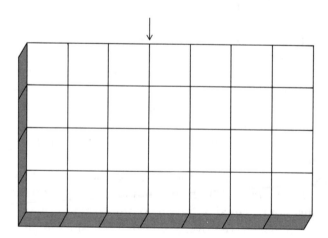

**Figure 32** *A format for Cutcake, with the first move being a vertical break at the arrow*

weight zero is dominated by some superstar. Only a feeling of incompleteness, Conway adds, prompts him to give a final theorem. Theorem 100 is: "This is the last theorem in this book."

## ANSWERS

The problem was to find the winning response to a first-player move in Cutcake. The cake is a 4 × 7 rectangle. If the first player breaks the cake vertically into a 4 × 4 square and a 4 × 3 rectangle, the unique winning reply is to break the 4 × 3 piece into two 2 × 3 rectangles.

## ADDENDUM

Since this chapter first ran as a *Scientific American* column in 1976, Conway has written other books and papers, and articles about him have appeared, some of which are listed in the bibliography. Especially notable is the two-volume *Winning Ways*, on which he collaborated with Elwyn Berlekamp and Richard Guy. It has already become a classic of recreational mathematics and has made important technical contributions to game theory and combinatorics. For Conway's work on Penrose tiling, see Chapters 1 and 2 of the book you now hold. For his work on sporadic groups and knot theory, see my *Scientific American* columns for June 1980 and September 1983, respectively. In 1987 Conway left Cambridge to accept a professorship at Princeton University.

Conway spoke on his game theories at a conference on recreational mathematics at Miami University in September 1976. His lecture, reprinted as "A Gamut of Game Theories" in *Mathematics Magazine* (see bibliography), concludes:

> The theories can be applied to hundreds and thousands of games—really lovely little things; you can invent more and more and more of them.
> It's especially delightful when you find a game that somebody's already considered and possibly not made much headway with, and you find you can just turn on one of these automatic theories and work out the value of something and say, "Ah! Right is 47/64ths of a move ahead, and so she wins."

## BIBLIOGRAPHY

"The G-Values of Various Games." Richard K. Guy and Cedric A. B. Smith, in *Proceedings of the Cambridge Philosophical Society*, 52, Part 3, 1956, pp. 514–526.

"A Mathematical Investigation of Games of 'Take Away'." Solomon W. Golomb, in *Journal of Combinatorial Theory*, 1, 1966, pp. 443–458.

"All Numbers Great and Small." John H. Conway. University of Calgary Mathematical Research Paper No. 149, 1972.

"All Games Bright and Beautiful." John H. Conway. University of Calgary Mathematical Research Paper No. 295, 1975. Reprinted in *The American Mathematical Monthly*, 84, 1977, pp. 417–434.

"A Gamut of Game Theories." John H. Conway, in *Mathematics Magazine*, 51, 1978, pp. 5–12.

*Winning Ways* (two volumes). Elwyn Berlekamp, John Conway, and Richard Guy. Academic Press, 1982.

### On Surreal Numbers

*Surreal Numbers*. Donald E. Knuth. Addison-Wesley, 1974.

*An Introduction to the Theory of Surreal Numbers*. Harry Gonshor. London Mathematical Society Lecture Note Series 110, 1986.

"Infinity Plus One and Other Surreal Numbers." Polly Shulman, in *Discover*, December 1995, pp. 97–105.

"The Man Who Played God with Infinity." Robert Matthews, in *New Scientist*, September 2, 1995, pp. 36–40.

### About Conway

"John Horton Conway: Mathematical Magus." Richard K. Guy, in *Two-Year College Mathematics Journal*, 13, 1982, pp. 290–299.

"John Horton Conway: A Mathematical Madness in Cambridge." Gary Taubes, in *Discover*, August 1984, pp. 40–50.

"Mathematical Mania." Mikael Thompson and Mark Niemann, in *Science Today*, Spring 1987, pp. 8–9.

"Intellectual Duel." Malcolm Browne, in *The New York Times*, August 30, 1988.

"The Things that Glitter." Caroline Moseley, in *Princeton Alumni Weekly*, December 23, 1992, pp. 10–13.

"At Home in the Elusive World of Mathematics." Gina Kolata, in *The New York Times*, October 12, 1993.

"John Horton Conway—Talking a Good Game." Don Albers, in *Mathematical Horizons*, Spring 1994, pp. 6–9.

"Mathematician." Charles Seife, in *The Sciences*, May/June 1994, pp. 12–15.

# CHAPTER 5

## Back from the Klondike and Other Problems

The "combinatorial revolution" in mathematics is still exploding as books and articles on combinatorics continue to proliferate. The computer is certainly contributing to the revolution by its power to analyze combinatorial problems that are too complex to be approached any other way. Another stimulating force is the increasing application of combinatorial theory to science and technology, particularly in particle physics and molecular biology. On a large scale the universe may be a collection of continuums to be handled by calculus, but on the microlevel it is a jumble of discrete elements with mysterious jumps and curious combinatorial properties. In some modern theories even time and space are quantized.

Hundreds of intriguing combinatorial puzzles, some old, some new, are now receiving the attention of serious mathematicians. We shall first take a look at an amusing prize puzzle devised by Sam Loyd that has recently been "cooked," or invalidated, by a computer program and then follow with several combinatorial problems that have no general solution in sight but present challenging tasks on their lower levels.

The greatest puzzle book ever published in the United States is a large volume, bound in pale green cloth, that has on its cover *Sam Loyd's Cyclopedia of 5000 Puzzles, Tricks and Conundrums, with Answers.* The spine bears a price of $5, but today one is lucky to find a copy for less than $25. Almost nothing is known about its publishing history. Although it is profusely illustrated, none of its several artists is identified. Copies bear the imprint of either the Morningside Press or the Lamb Publishing Company, both of New York, but no one knows which edition came first.

All copies are dated 1914. Because this was three years after Loyd's death, it was widely assumed that his son, who took his father's name, had selected the puzzles from older newspapers and magazines and had patched them hastily into one massive volume. (The book is peppered with omissions, mistakes and printer's errors.) It was later discovered that the elder Loyd had edited and published a quarterly called *Our Puzzle Magazine,* which first appeared in June 1907. Copies of it are now exceedingly rare. The *Cyclopedia* is simply a reprinting of this magazine made from its unaltered page plates.

Loyd's "Back from the Klondike" puzzle, shown in Figure 33, appeared in the second issue (October 1907) of his magazine and is re-

**Figure 33** *Sam Loyd's solution to his Klondike puzzle*

printed on page 106 of the *Cyclopedia*. It is not known whether Loyd called it a "prize puzzle" because a prize for its solution was offered to readers of the magazine or whether it had been a contest puzzle in some earlier incarnation.

"Start from that heart in the center," Loyd says, "and go three steps in a straight line in any one of the eight directions, north, south, east or west, or on the bias, as the ladies say, northeast, northwest, southeast or southwest. When you have gone three steps in a straight line, you will reach a square with a number on it, which indicates the second day's journey, as many steps as it tells, in a straight line in any one of the eight directions. From this new point when reached, march on again according to the number indicated, and continue on, following the requirements of the numbers reached, until you come upon a square with a number which will carry you just one step beyond the border, when you are supposed to be out of the woods and can holler all you want, as you will have solved the puzzle."

Loyd's sneaky solution is shown by the curved black line. The path simply shuttles up and down along a main diagonal until a number is reached that makes possible what Loyd, in his answer, calls a "bold strike via S.E. to liberty!" The solution, Loyd claims, is unique. "Those who failed to master it readily discovered that one false step at any stage of the game throws one into the whirlpool from which there is no egress."

But the old maestro was wrong! Early in 1976 Penelope J. Greene, a graduate student in sociology at the University of Washington, R. Duncan Mitchell, a graduate student in economics there, and Horace A. Greene, a science teacher at Roosevelt University in Chicago, made a combined attack on the problem with a Fortran program written by Mitchell. In a few seconds the program found Loyd's solution along with an explosion of hundreds of others. Indeed, one can leave the center 3 in any of the eight directions and escape from the woods. The start of one alternate route is shown in Figure 34. All these other routes lead eventually to 4, a key link on the diagonal, from which the path is completed as it is in Loyd's solution.

The three programmers noticed something unusual about the alternate answers. One cell and one only, not a link in Loyd's solution, is common to all the alternative routes. They called it the "troublesome 2" (circled in illustration). Could this have been an artist's error? It seems likely, because when the programmers substituted other digits for 2, the computer found only one nonzero digit that eliminates all alternative paths.

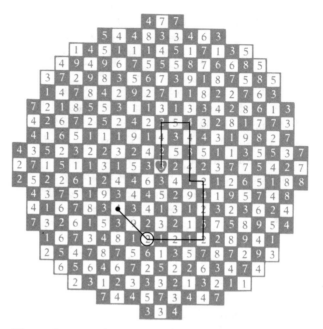

**Figure 34** *An alternate path through the "troublesome 2"*

This is the pleasant task — it is really quite easy — that I offer readers. Correct Loyd's Klondike puzzle by changing the troublesome 2 to the only nonzero digit that restores uniqueness to Loyd's solution. In calling Loyd's solution unique, we discount wasted "side trips," where one leaves a cell only to return to it again before proceeding or where one leaves cell *A* for a side trip to *B* when one can go directly from *A* to *B*.

The origin of Chinese checkers, a board game popular around the world since about 1920, is obscure. Apparently it was first marketed under this trade name in the United States. It has no connection with China and almost no resemblance to checkers, but the name has persisted.

The standard board is shown in Figure 35. The holes are numbered for convenience in recording moves, and borders have been added to outline all star boards of smaller size. If two people play, 10 marbles of one color are placed in the triangular "yard" formed by holes 1 through 10 and 10 marbles of another color go in holes 112 through 121. Rules of play are similar to those of Halma (see Chapter 11 of my *Wheels, Life, and Other Mathematical Amusements*, W. H. Freeman and Company, 1983), a popular British game that probably suggested the variant represented by Chinese checkers.

Two kinds of move are allowed. A "step" is a move in any of six directions to an adjacent vacant hole. A "hop" is a move over an adjacent marble, again in any of the six directions, to a vacant hole on the other side. The hop is like a checker jump except that the jumped piece is not removed and can be either one's own piece or an opponent's. A piece can continue hopping as long as possible or the chain can be stopped at any point. Hops are never compulsory. The combining of steps and hops

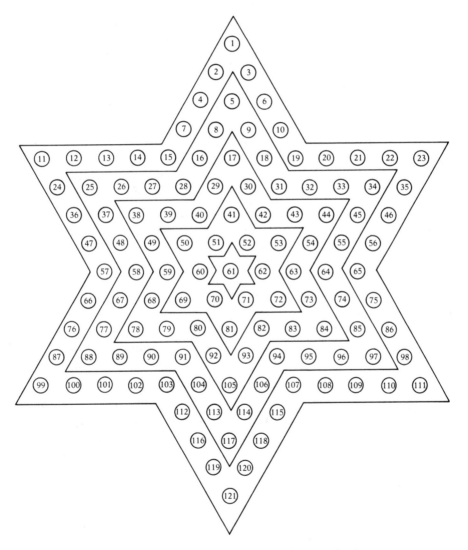

**Figure 35** *A variety of boards for Chinese checkers*

is forbidden. The winner is the first player to occupy all the holes of his opponent's starting yard.

The game is obviously unplayable on the degenerate smallest "star" of one hole. On the next larger star each player has one marble and the game is a trivial win for the first player to move. The increase in complexity as the size of the board is increased to three marbles on each side is enormous. I believe it is not known which player has the win.

On this star the game is quite pleasant. To prevent one player from forcing a draw by keeping a marble perpetually in its home yard one more rule must be added: If a counter can leave its yard by hopping an enemy counter, it must do so. Once it is out of its yard, it may not return, although it is allowed to pass *through* the yard in a chain of hops.

Let us consider on each star board the following solitaire problem. Marbles of one color are placed in starting positions on the bottom yard. What is the minimum number of moves (counting a chain of hops as a single move) necessary to transfer all of them to the yard at the top?

The problem is meaningless on the smallest star and trivial on the 13-hole star. For the star with a yard of three marbles 11 moves will do it. Example: 92–81, 81–71, 105–82–61, 61–52, 93–82, 82–61–42, 42–30, 71–61, 61–42–17, 52–41, 41–29.

For the six-marble star the best solution I have been able to get is 18 moves. Starting with 113–93 there are many ways to construct in nine moves a vertical "ladder" of marbles on holes 9, 30, 52, 71, 93 and 114. Nine more moves, reversing the previous play, put the six marbles in the top yard.

Now for our main problem: What is the minimum number of moves needed to transfer 10 marbles from one yard to the opposite yard on the standard Chinese checkers board? I first heard of this problem in 1961 from Octave Levenspiel, professor of chemical engineering at Oregon State University. He wrote that the best he could do was 31 moves but that his mother, "the family puzzle solver, chess and bridge champ," had once solved it in 28 but had failed to record the moves. *Ibidem*, a Canadian periodical on magic (now defunct), published in August 1969, a "proof" that 29 is the minimum. Then in 1971 Levenspiel, trying to reconstruct his mother's 28-move solution, hit on a spectacular solution in 27! Harry O. Davis of Portland, OR., believes he has a proof that 27 is the minimum, but the proposition has not yet been confirmed.

Can the reader find a 27-move solution? Chinese checkers is sometimes sold with a larger board that has 25 marbles in each yard. For such a board the best solution in my files, sent to me in 1974 by Min-Wen Du of Taiwan, is 35 moves.

In the London *Tribune* for November 7, 1906, Henry Ernest Dudeney, Loyd's British counterpart, published the problem of placing 16 pawns on a chessboard so that no three are in line in any direction. (The problem later appeared as No. 317 in his *Amusements in Mathematics*.) By "line" is meant *any* straight line, not just orthogonals and diagonals. The pawns represent points at the centers of cells. Since then the generalized no-three-in-line problem, as it came to be known, has been the topic of several technical papers.

The main problem, which is still unsolved, is to answer the following question: On a square checkerboard of $n$ cells on a side ($n$ greater than 1) is it always possible to place $2n$ counters so that no three are in line? The old "pigeonhole" principle proves that $2n$ cannot be exceeded. No row or column can hold more than two counters, and since $2n$ counters require two in each orthogonal, one more counter is certain to put three counters in one row as well as three in one column. A less trivial argument (developed by R. R. Hall, T. H. Jackson, A. Sudbery and K. Wild in their 1975 paper) proves that at least $n$ counters can always be placed. For large boards these authors show that one can get quite close to $3n/2$ counters.

Michael A. Adena, Derek A. Holton and Patrick A. Kelly, in their 1974 paper, reported on computer programs that found all distinct solutions for $n$ equals 2 through 10. Rotations and reflections are excluded. The number of solutions are respectively 1, 1, 4, 5, 11, 22, 57, 51 and 156. Figure 36 gives an example for each $n$ from 2 through 10. Note the startling simplicity and symmetry of the order-8 solution!

At the time these authors wrote, no solution for $n = 11$ was known. They gave a solution for $n = 12$, but it proved to be invalid. They had failed to notice two rows of three in line.

Solutions for $n = 11$ and $n = 12$ were found in 1975 by D. Craggs and R. Hughes-Jones of the University of Kent and published in 1976. They are shown in Figure 37. Craggs and Hughes-Jones found five other solutions for $n = 11$ and three others for $n = 12$. The total number of solutions for these two values of $n$ is not known, and no one has constructed a solution for a square of an order higher than 12. Richard K. Guy and Patrick A. Kelly, in their 1968 paper, give arguments to support their conjecture that the number of orders with $2n$ solutions is finite. The smallest board on which a $2n$ solution is impossible is not known.

If we narrow the definition of "line" to an orthogonal or diagonal row, the problem is solved. The maximum cannot be greater than $2n$, and $2n$ is possible on all boards of orders higher than 1. For $n$ greater than 3 the problem is solved by superposing any two solutions of the

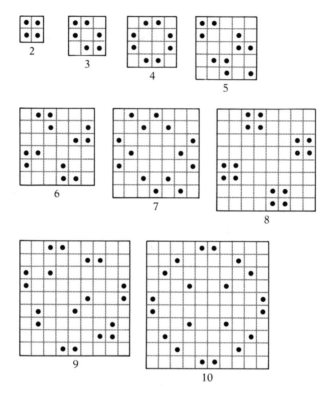

**Figure 36**  Some solutions to the maximum "No-Three-in-Line Problem"

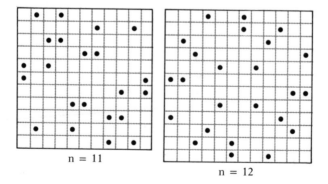

**Figure 37**  Recent solutions to the problem

classic nonattacking-queens problem. (This is the problem of placing $n$ queens so that no queen attacks another.) It is always possible to do this with a pair of solutions that put the $2n$ queens on $2n$ cells.

Instead of asking for the maximum number of counters that can be put on an order-$n$ board, no three in line, let us ask for the minimum that can be placed such that adding one more counter on any vacant cell will produce three in line. This intriguing problem has not yet received much attention from the experts.

If "line" is taken in the broadest sense—a straight line of any orientation—the problem is difficult. Figure 38 shows minimal solutions for $n$ equals 3 through 10. The sequence of counters is 4, 4, 6, 6, 8, 8, 8, 10. The last three patterns were found by the British statistician Stephen Ainley and included in his book *Mathematical Puzzles* (G. Bell and Sons, 1977). The pattern for order 10 also suffices for order 11. Ainley found 12-counter solutions for orders 12, 13 and 14.

The problem is also unsolved if "line" is restricted to orthogonals and diagonals. In other words, what is the smallest number of pawns you can put on a board of side $n$ such that no pawn can be added without creating three in a row, a column or a diagonal? Another way to view the problem is to see it as a game in which two players alternately place a counter on a square board until one player loses by being forced to make three in an orthogonal or diagonal line. How short can such a game be? A puzzle asking for a solution on the order-8 board appeared in *Technology Review* (June 1967) in Allan J. Gottlieb's excellent "Puzzle Corner." This is the only publication known to me of a problem of this type.

The solutions shown in Figure 38 for orders 3 through 7 are also solutions for the minimal problem when "line" is defined narrowly. Minimal solutions for orders 8, 9 and 10 are shown in Figure 39. The last pattern also solves orders 11 and 12. These are far from unique patterns. There are dozens of solutions for $n = 8$, many with bilateral symmetry and others with twofold symmetry. If 10 is minimal, fourfold symmetry is ruled out because 10 is not evenly divisible by 4.

## ANSWERS

Sam Loyd's solution to his Klondike puzzle is unique if the "troublesome 2" is changed to 1. It is easy to show that any other nonzero digit in this cell provides an alternate escape route. Digits 4, 5, 6, 8 and 9 lead

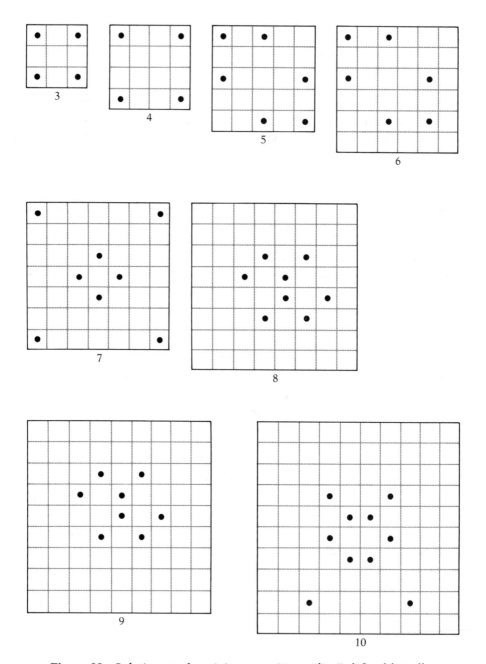

**Figure 38** Solutions to the minimum problem, "line" defined broadly

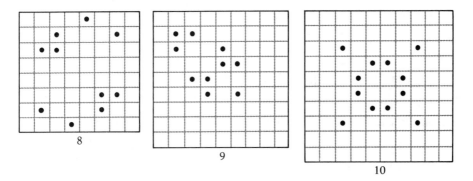

**Figure 39** *Solutions to minimum problem, "line" defined narrowly*

immediately to escape, 3 provides escape in two steps to the southwest and 7 offers escape in two steps due west.

Ten marbles on a Chinese checkers board can be transferred to the opposite yard in 27 moves:

1. 115–106
2. 120–115–93
3. 116–105–82
4. 82–72
5. 118–105–82–63
6. 121–116–105–82
7. 114–94–71–73–53
8. 53–42
9. 112–114–94–71–73–53–30
10. 119–114–94–71–73–53
11. 106–81–83–62–43–41–18
12. 18–9
13. 113–114
14. 117–106–81–83–62–43–41–18–5
15. 9–8
16. 114–106
17. 106–81–83–62–43–41–18
18. 72–54–52–31–9–2

19. $93 - 72 - 54 - 52 - 31 - 9 - 7$
20. $82 - 72$
21. $72 - 54 - 52 - 31 - 9$
22. $42 - 17 - 6 - 4 - 1$
23. $63 - 42 - 17 - 6$
24. $53 - 42$
25. $42 - 17 - 4$
26. $30 - 10 - 3$
27. $18 - 10$

The solution is palindromic: the second half is a mirror image of the first.

## ADDENDUM

Numerous readers pointed out that Sam Loyd's solution to his Klondike puzzle has an alternate last move of the same length: A "bold strike" northwest also leads to freedom. Harold F. Bennett reminded me that he had written to me in 1966 to say that by using an ingenious hand technique he had found scores of alternate solutions. I had filed his letter and forgotten about it. Maxim G. Smith also sent an excellent method of finding other paths without computer help.

All alternate paths eventually reach a 4-cell that can be reached from the center in one move. It follows that Loyd's solution is the shortest. Perhaps Loyd had this in mind when he said his solution was unique.

Will Shortz, a senior editor of *Games* magazine, located the first publication of the Klondike puzzle in the *New York Journal and Advertiser*, April 24, 1898, where it occupied a full page. Two cells bear numbers that differ from those in Loyd's *Cyclopedia* version. In row seven the tenth number from the left is 2, and in row nine the sixth number is 8. The *Journal* offered $10 for the "best answer received within two weeks." "Best," one assumes, meant shortest. In his answer (May 15) Loyd said 10,000 letters had been received, but "few of the clever pathfinders discovered the shortcut of seventeen steps."

Unfortunately for Loyd, he overlooked the consequence of 8 in the cell mentioned above. A "Mr. Cauler" of Plainfield, N.J., won the prize with just five moves: 3 south, 1 northwest, 4 northwest, 1 east and 8 north. To eliminate this solution Loyd later changed the 8 to 1. I don't know why he changed the other cell from 2 to 3. Perhaps the 2 allowed still another unintended solution.

Virginia F. Walters, Octave Levenspiel and Mrs. M. Reynolds lowered the record for a transfer of six marbles on the order-4 Chinese checkers board from 18 to 17 moves. On these smaller boards you are not, of course, allowed to use holes beyond the borders if you are working on a standard-size board.

The no-three-in-line problem brought a flood of letters. Many readers misinterpreted the problem, overlooking the statement that

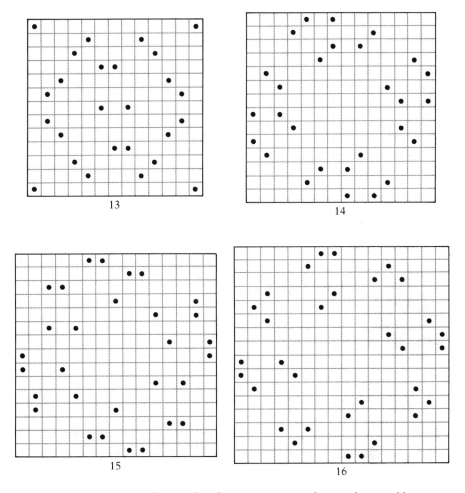

**Figure 40** *New solutions for the maximum no-three-in-line problem*

"line" is not confined to rows, columns or diagonals. Among those who understood the task, significant advances were made.

Richard Byfield, Richard Jacobson, Anne De Lamper and Robert Van Clampitt each found by hand an order-13 solution. Michael Meierruth wrote a nonexhaustive computer program that turned out 29 solutions of order 13 and one of order 15. Later he found a solution for order 14 and four solutions for order 16. Eric Jamin found by hand four order-14 solutions and one for order 16. Figure 40 shows sample patterns for 13, 14 and 16 and the only known pattern for 15.

Solutions are now known for orders through 32. For a discussion of the problem, with some interesting conjectures and a bibliography, see Problem F4 in Richard Guy's *Unsolved Problems in Number Theory*.

On the minimum no-three-in-line problem with lines of the queen type let $n$ be the order of the square and $p$ the minimal number of counters. John Harris was able to show that $p$ is at least $n$ except when $n = 3$ (modulo 4), in which case $p$ could be $n - 1$. Can $p$ be odd? As far as I know, this has not been answered.

B.C. by Johnny Hart  © 1988, Creators Syndicate, Inc.

*"I've noticed that no three are alike"*

# BIBLIOGRAPHY

*Amusements in Mathematics.* Henry E. Dudeney. Dover, 1958.

*The Best Mathematical Puzzles of Sam Loyd.* Dover, 1959.

*More Mathematical Puzzles of Sam Loyd.* Dover, 1960.

"The No-Three-in-Line Problem." Richard K. Guy and Patrick A. Kelly, in *Canadian Mathematics Bulletin*, 11, 1968, pp. 527–531.

"Some Thoughts on the No-Three-in-Line Problem." Michael A. Adena, Derek A. Holton and Patrick A. Kelly, in *Combinatorial Mathematics: Proceedings of the Second Australian Conference*, edited by Derek A. Holton, 403 of *Lecture Notes in Mathematics*, Springer-Verlag, Berlin, 1974.

"Some Advances in the No-Three-in-Line Problem." R. R. Hall, T. H. Jackson, A. Sudbery and K. Wild, in *Journal of Combinatorial Theory*, 18, 1975, pp. 336–341.

"On the No-Three-in-Line Problem." D. Craggs and R. Hughes-Jones, in *Journal of Combinatorial Theory*, 20, 1976, pp. 363–364.

"On the No-Three-in-Line Problem, II." Torleiv Klove, in *Journal of Combinatorial Theory*, 24, 1978, pp. 126–127.

"On the No-Three-in-Line Problem, III." Torleiv Klove, in *Journal of Combinatorial Theory*, 26, 1979, pp. 82–83.

"Update on the No-Three-in-Line Problem." David Brent Anderson, in *Journal of Combinatorial Theory*, 27, 1979, pp. 365–366.

*Unsolved Problems in Number Theory.* Richard Guy. Springer-Verlag, 1981, Problem F4.

"No-Three-in-Line for Seventeen and Nineteen." H. Harborth, P. Oertel, and T. Prellberg, in *Discrete Mathematics*, 73, 1988/89, pp. 89–90.

"Progress in the No-Three-in-Line Problem." Achim Flammenkamp, in the *Journal of Combinatorial Theory*, Series A, Vol. 60, 1992, pp. 305–311. The author gives solutions up to $46 \times 46$ boards.

# CHAPTER 6
## The Oulipo

> Rhodes roams leads to all?
> Roads lead all to Rome,
> Lead all Rhodes to Rome.
> Rome-roads lead to all?
> All roam roads to Leeds!
> Rome rode all to Leeds.

*Finnegans Wake* continues to be the towering example of how serious literary intent can be combined with outrageous wordplay. Other notable instances are e e cummings' eccentric typography, Gertrude Stein's nonsense verse and the constant wordplay in Vladimir Nabokov's fiction.

For the past 15 years the most sophisticated and amusing examples of literary wordplay have been produced by a whimsical, slightly mad French group called the Oulipo. The name comes from Ouvroir de Littérature Potentielle (Workshop of Potential Literature). Although this chapter will be mostly about the Oulipo, I shall digress frequently to cite the work of comparable experts in English.

Most of what follows derives from three sources: (1) *La Littérature Potentielle (Créations Re-créations Récreations)*, a paperback by the Oulipo (Éditions Gallimard, 1973); (2) a marvelous article on the Oulipo by its only American member, Harry Mathews, which appeared in *Word Ways*, a quarterly journal of recreational linguistics, edited and published by A. Ross Eckler, obtainable from Eckler, Spring Valley Road, Morristown, NJ 07960; and (3) correspondence with Mathews, who lives part of the time in Paris.

The Oulipo was founded in 1960 by two brilliant Frenchmen: François Le Lionnais and the late Raymond Queneau. Le Lionnais is a mathematician and author of many books, including six on chess. Queneau, who died in 1976 at the age of 73, was one of France's most influential writers. He is best known for his 1959 novel *Zazie dans le Métro*, about the adventures of an 11-year-old Parisian nymphet. (It was made into a popular motion picture by Louis Malle.) Another of Queneau's well-known works, *Exercises in Style*, consists of a trivial event—a man on a bus is pushed and later talks to a friend about adding a button to his overcoat—told in 99 different styles, one of them entirely in pig latin. Queneau studied mathematics in his youth. It remained a major interest that permeated his novels, poetry and criticism. For a good account of his work by John Updike, another writer knowledgeable about mathematics and given to wordplay, see Updike's *Hugging the Shore*, pages 398–409.

The other French members of the Oulipo are, in alphabetical order, Noël Arnaud, Marcel Bénabou, Jacques Bens, Claude Berge, Paul Braffort, Jacques Duchâteau, Luc Étienne, Paul Fournel, Jean Lescure, Michèle Métail (the only woman in the group), Georges Perec, Jean Queval and Jacques Roubaud. The foreign members are André Blavier (Belgium), Italo Calvino (Italy) and Mathews. All members are mathematicians or writers or both.

Mathews was born in New York City in 1930. He was graduated from Harvard College in 1952 with a degree in music and has been living in Europe ever since. Apart from books of poetry he has written three wild and funny novels: *The Conversions* (Random House, 1962), *Tlooth* (Doubleday, 1966) and *The Sinking of the Odradek Stadium and Other Novels* (Harper & Row, 1975). The "other novels" are reprints of the two earlier ones. All three books are filled with quasi-Oulipian wordplay, notably a four-page scene in *Tlooth* where the pornography is obscured by continuous spoonerisms of mounting combinatorial complexity.

*Tlooth* is a picaresque tale of blue, blasphemous, black humor. Nephthys Mary Allant is the bisexual catcher of the Baptist baseball team at Jacksongrad a Siberian labor camp. We do not learn her name or sex until 10 pages from the novel's end. (It is not until the very end of *Conversions* that the reader discovers that the narrator is a mulatto.) Mary had been a violinist until her former friend Evelyn Roak, a surgeon, needlessly amputated two fingers of her left hand. Dr. Roak is also an inmate of the camp, having been sent there for selling her amputations to a delicatessen. She plays on a rival ball team. Mary tries unsuccessfully to kill her. An infection in Mary's finger stumps prevents her from practicing dentistry, the profession for which she is in training.

Evelyn is released from the camp, and Mary later escapes. After wandering though Asia and Europe looking for Evelyn, Mary (her infection now cured by folk medicine) opens a dental office near Lyons. Evelyn shows up as a patient, Mary's finger bones dangling from her charm bracelet. Mary intends to finish her off in the chair, but Evelyn's gums disclose that she is dying of smallpox. Mary gives her medication and sends her off "to her stars."

The book's title is an enigmatic word gasped by the warm mud of an oracular bog after Mary completed the ritual of inserting her bare leg in the ooze for 68 seconds at vespers, then withdrawing it quickly. The word has overtones of "Tooth," "Truth" and Tlooth Lautrec. Reading the book, said one critic, is "like riding a bumper car through a dictionary whose definitions are askew."

Mathews is currently writing his fourth novel, which is based on what the Oulipo calls "Mathews' algorithm." In this procedure the elements of a work are cyclically permuted in two ways to form two matrixes, which are then read by taking the columns of one matrix from the top down and the columns of the other matrix from the bottom up. The elements can be letters, words, sentences, paragraphs, chapters, ideas— anything you like. Unfortunately, there are no good examples in English.

Mathews' first contribution to *Word Ways* was a list of common words that are spelled the same in both French and English but have entirely distinct meanings. This short dictionary of "L'Égal Franglais" was compiled to aid Oulipo members in producing ambivalent Anglo-French texts. Mathews' entire list, which was not meant to be exhaustive, is presented in Figure 41 for readers who would like to experiment with sentences that have one meaning in English and another in French.

The Oulipo's first manifesto was Queneau's *Cent Mille Milliards de Poèmes (A Hundred Thousand Milliard Poems)*, published by Éditions Gallimard, where Queneau was a senior editor. The book consists of 10 basic sonnets, each one on a right-hand page that is sliced into 14 strips (a strip for each sonnet line). By flipping strips to the left or right one can obtain $10^{14}$ (100,000 billion) sonnets, all structurally perfect and making sense. (Mechanical books for children have employed the same format to produce pictorial combinations of parts of animals or people, sometimes also scrambling syllables of vertically printed names to get such bizarre beasts as the elepotamus and the kangaboon.) Flip the strips and you can read a sonnet that (probably) no one has read before and no one will ever read again.

The lipogram, an ancient type of wordplay, is a sentence or a longer work that omits one letter or more of the alphabet. The major specimen

| | | | | | |
|---|---|---|---|---|---|
| a | cause | fat | lecture | palace | rude |
| ache | caution | fee | l'egal | pan | |
| ail | chair | fend | legs | pane | sale |
| allege | champ | fin | Lent | par | sang |
| amends | chat | fit | lice | pare | saucer |
| an | choir | fond | lie | pat | scare |
| ante | chose | font | l'imitation | pate | signet |
| appoint | coin | for | l'ion | pays | singe |
| are | collier | forage | lit | pester | son |
| as | comment | fore | l'izard | pet | sort |
| at one | con | fort | location | Peter | sot |
| attend | confection | four | loin | pie | spire |
| audit | corner | fur | longer | pied | stage |
| averse | cote | | l'oser | pin | stance |
| axe | courtier | gale | love | pincer | store |
| | crane | gate | | pine | sue |
| Bade | crisper | gave | ma | plains | suit |
| bail | cure | gaze | mail | plate | super |
| ballot | | gene | main | plier | supplier |
| barber | | gent | mange | plies | |
| bard | dam | gourd | manger | plot | tale |
| baste | D'ane | gout | mare | pour | tape |
| bat | d'are | grief | mariner | pries | tenant |
| be | d'art | grime | men | prone | the |
| beat | defiance | groin | mien | | these |
| bee | defile | guise | mince | rang | tiers |
| Ben | dent | | mire | range | tin |
| bide | derive | hair | miser | ranger | tine |
| bled | design | hale | moult | rape | tint |
| blinder | d'etain | harder | mute | rate | tire |
| bond | dime | hate | | rave | ton |
| bore | dire | have | n'est | rayon | toper |
| borne | dive | here | net | rebut | tort |
| bout | don | hurler | Nil | reel | tot |
| bribes | don't | | noise | regain | tout |
| bride | dot | if | n'ose | regal | tries |
| but | dresser | | n'ote | rein | |
| butter | drill | jars | | relent | van |
| | d'un | | oil | rend | vent |
| | d'une | labour | on | report | venue |
| can | | lad | once | ride | verge |
| cane | edit | laid | or | ripe | verse |
| canner | emu | l'air | ours | river | vie |
| cap | engraver | lame | | robin | viol |
| car | enter | l'ane | pain | rogue | |
| carrier | entrain | layer | pair | Roman | |
| carter | ere | lecher | pal | rot | |
| case | fade | | | | |

*Figure 41* *Harry Mathews' short dictionary of "L'Égal Franglais"*

in English is *Gadsby*, a novel by Ernest Vincent Wright that was published in Los Angeles in 1939. It does not contain a single *e*, the letter that appears most frequently in both English and French.

Experiments with lipograms by the Oulipo culminated in Georges Perec's novel *La Disparition (The Disappearance)*, published in 1969 by

Denoël, Les Lettres Nouvelles. In it too *e* is the disappearing letter. The novel is much longer and better than *Gadsby*. Mathews describes it as an "elaborate, funny story of unbelievable virtuosity," so well written that some critics praised it without noticing anything strange! Perec considered following it with another novel in which *e* is the *only* vowel.

Perec is currently working on a novel called *La Vie, Mode d'Emploi (Life and How to Use It)* that is based on, among other things, a Greco-Latin Square of order 10. The great Leonhard Euler had conjectured that such a square could not exist, but one was found in 1959, and it provided a colorful cover for the November issue of *Scientific American* that year. Although Perec is a prolific French writer, only his first book, *Les Choses: Une Histoire des Années Soixant*, has been published in the United States.

Perec is also responsible for the Oulipo's longest palindrome. It is about palindromes and contains more than 5,000 letters that begin and end:

> *Trace l'inégal palindrome. Neige. Bagatelle, dira Hercule. Le brut repentir, cet écrit né Perec. L'arc lu pèse trop, lis à vice-versa. . . . Désire ce trépas rêvé: Ci va! S'il porte, Sépulcral, ce repentir, cet écrit ne perturbe le lucre: Haridelle, ta gabegie ne mord ni la plage ni l'écart.*

Mathews translates this passage as follows:

> Trace the unequal palindrome. Snow. A trifle, Hercules would say. Rough penitence, this writing born as Perec. The read arch is too heavy: read vice versa. . . . Desire this dreamed-of decease: Here goes! If he carries, entombed, this penitence, this writing will disturb no lucre: Old witch, your treachery will bite into neither the shore nor the space between.

The most skillful composer of English palindromes (in my opinion he is also England's best writer of comic verse) is J. A. Lindon. *Word Ways* has published many of his palindromic poems; others can be found in Howard W. Bergerson's *Palindromes and Anagrams*. Lindon's finest palindromic achievement, a dialogue about the seduction of Eve, appears in Figure 42. Every line is a letter-unit palindrome, and so is the title.

A famous Oulipo algorithm called $S + 7$ (for "Substantif plus 7") was invented by Lescure. In English it is $N + 7$. The procedure is to replace each noun in a familiar prose passage with the seventh noun that follows it in a specified dictionary. $N + 7$ is a special case of the more general algorithm $M \pm n$, where $M$ is any kind of *mot* (word) and $n$ is any positive integer. In both the text and the dictionary unhyphenated compound words, such as "high school," are ignored. Here, for instance, are the

In Eden, I

ADAM: Madam —
EVE: Oh, who —
ADAM: (No girl-rig on!)
EVE: Heh?
ADAM: Madam, I'm Adam.
EVE: Name of a foeman?
ADAM: O stone me! Not so.
EVE: Mad! A maid I am, Adam.
ADAM: Pure, eh? Called Ella? Cheer up.
EVE: Eve, not Ella. Brat-star ballet on? Eve.
ADAM: Eve?
EVE: Eve, maiden name. Both sad in Eden? I dash to be manned, I am Eve.
ADAM: Eve. Drowsy baby's word. Eve.
EVE: Mad! A gift. I fit fig, Adam . . .
ADAM: On, hostess? Ugh! Gussets? Oh, no!
EVE: ???
ADAM: Sleepy baby *peels.*
EVE: Wolf! Low!
ADAM: Wolf? Fun, so snuff "low."
EVE: Yes, low! Yes, nil on, no linsey-wolsey!
ADAM: Madam, I'm *Adam.*
　　*Named under a ban.*
　　*A bared, nude man —*
　　　　　　　　　　　Aha!
EVE: Mad Adam!
ADAM: Mmmmmmmm!
EVE: Mmmmmmmm!
ADAM: Even in Eden I win Eden in Eve.
EVE: Pure woman in Eden, I win Eden in — a mower-up!
ADAM: Mmmmmmmm!
EVE: Adam, I'm Ada!
ADAM: Miss, I'm Cain, a monomaniac. Miss, I'm —
EVE: No, son.

**Figure 42** *A palindromic dialogue by J. A. Lindon*

first two sentences of *Moby Dick*, which I have altered $N + 7$ by using *Webster's New Collegiate Dictionary*:

> Call me islander. Some yeggs ago — never mind how long precisely — having little or no Mongol in my purulence, and nothing particular to interest me on shortbread, I thought I would sail about a little and see the watery partiality of the worriment.

Another Lescure algorithm is to reverse the order of a given type of word, switching first and last instances, then the second from each end

ADAM: Name's Abel, a male base man.
EVE: Name not so, O Stone man!
ADAM: Mad as it is it is Adam.
EVE: I'm a Madam Adam, am I?
ADAM: Eve?
EVE: Eve mine. Denied, a jade in Eden, I'm Eve.
ADAM: No fig. (Nor wrong if on!)
EVE: ???
ADAM: A daffodil I doff, Ada.
EVE: 'Tis a—what—ah, was it—
ADAM: Sun ever! A bare Venus . . .
EVE: 'S pity! So red, ungirt, rig-nude, rosy tips . . .
ADAM: Eve is a sieve!
EVE: Tut-tut!
ADAM: Now a see-saw on . . .
EVE: On me? (O poem!) No!
ADAM: Aha!
EVE: I won't! O not now, I—
ADAM: Aha!
EVE: NO! O God, I—(Fit if I do?) *Go on.*
ADAM: Hrrrrrrh!
EVE: Wow! Ow!
ADAM: Sores? (Alas, Eros!)
EVE: No, none. My hero! More hymen, on, on . . .
ADAM: Hrrrrrrrrrrrrrrrrrrrrrrh!
EVE: Revolting is error. Resign it, lover.
ADAM: No, not now I won't. On, on . . .
EVE: Rise, sir.
ADAM: Dewy dale, cinema-game . . . Nice lady wed?
EVE: Marry an Ayr ram!
ADAM: Rail on, O liar!
EVE: Live devil!
ADAM: Diamond-eyed no-maid!
BOTH: Mmmmmmmmmmmmmmmmmmmm!

and so on. Applied to nouns in the first chapter of *Moby Dick*, the opening sentences are:

> Call me air. Some hills ago—never mind how long precisely—having little or no phantoms in my whale, and nothing particular to interest me on processions, I thought I would sail about a little and see the watery soul of the purpose.

Queneau's booklet *Les Fondements de la Littérature* transforms David Hilbert's axioms for Euclidean geometry by replacing the words "points," "lines" and "planes" respectively with "words," "phrases" and "paragraphs" to obtain a new set of axioms for which Queneau provides a witty commentary. Another Oulipo algorithm is to replace words with dictionary definitions or with freer definitions of the kind found in crossword puzzles. The new statement is transformed again by the same method until all original meaning is lost. A special challenge is to begin with two statements that have entirely different meanings and then transform them in the fewest number of steps until they are identical.

Joining the first half of one proverb to the second half of another produces the "perverb," a form explored by Mathews in both French and English. His *Selected Declarations of Dependence* (Eternal Network, Toronto, 1976) consists of poems and prose pieces that explore hundreds of perverbs derived from 46 English proverbs. Here are some examples:

"A rolling stone gets the worm."

"A bird in the hand waits for no man."

"The road to hell has a silver lining."

"It's an ill wind that spoils the broth."

One may also modify a proverb by substituting homophones — words that sound much the same but have different meanings. Two classic English examples are "There's no fuel like an oil fuel" and "There's no police like Holmes." Proverbs can also be changed in startling ways simply by truncating them:

"People who live in glass houses shouldn't."

"All work and no play makes jack."

"Familiarity breeds."

Lescure has experimented with sentences containing just four principal words by testing all 24 permutations. The goal is to maximize the number of permutations that make sense. Homophones are allowed. An English example by Mathews is shown in Figure 43. The list is incomplete, he points out, because the remaining permutations are redundant.

Lines of poems are permuted by the Oulipo in many carefully defined ways. Luc Étienne uses a Möbius strip to interweave the lines of a poem to make a different poem. Write half of the poem on one side of a strip and the other half (upside down) on the opposite side. Twist and join the ends. Reading around the single side of the Möbius band inter-

A Partial Survey of Western European Holiday Migrations

EXODUS A

Leeds' roads roam to all?
Rome's Leeds' road to all—
All Leeds rode to Rome.

EXODUS B

All Rome leads to roads.
Rome leads all to roads,
Leads Rome all to roads:
"Roam all leads to roads!"
Roads lead Rome to all?
All leads roam to Rhodes,
Lead all Rome to Rhodes.

RETURN A + B

Rhodes roams leads to all?
Roads lead all to Rome,
Lead all Rhodes to Rome.
Rome-roads lead to all?
All roam roads to Leeds!
Rome rode all to Leeds.

SUMMARY

All roads roam to Leeds

*Figure 43*   *Mathews' permutations of a proverb*

laces the lines in the way that clasping your hands interlaces your fingers. The goal is to compose poems that have opposite meanings when they are read before and after the twist.

The Möbius construction can be applied to two stanzas of the same length by different poets. The resulting poem is usually grotesque, but not always. In Figure 44 I have employed the Möbius transformation to combine Edna St. Vincent Millay's famous quatrain about candleburning with the first quatrain of Elinor Wylie's beautiful lyric "Bells in the Rain." The composite poem can be read by starting either with Millay's quatrain or with Wylie's.

Sleep Falls
by Elinor Millay

Sleep falls, with limpid drops of rain
(My candle burns at both ends)
upon the steep cliffs of the town.
It will not last the night.

Sleep falls, men are at peace again.
But ah, my foes and oh, my friends,
while the small drops fall softly down,
It gives a lovely light.

My Candle
by Edna Wylie

My candle burns at both ends.
Sleep falls with limpid drops of rain
(it will not last the night)
upon the steep cliffs of the town.

But ah, my foes and oh, my friends,
sleep falls, men are at peace again.
It gives a lovely light
while the small drops fall softly down.

*Figure 44*   *Interwoven poems for a Möbius strip*

Oulipo members have written short stories and plays with junctures at which the reader or audience can choose between alternate transitions to obtain different plot combinations. (This technique should not be confused with the randomizing of fictional elements, as in the boring examples of Michel Butor's *Mobile* or the recent novels of William S. Burroughs.) *Hopscotch*, a novel by the Argentine expatriate (he too lives in Paris) Julio Cortázar, is closer to the Oulipo intent. Its 154 chapters are designed to be read in two ways. First you read conventionally from Chapter 1 through Chapter 56. Then you start at Chapter 73 and hopscotch through the book by taking chapters in a sequence given by the number at the end of each chapter. Many chapters are read twice, and one is read four times. A translation from the Spanish was published by Pantheon in 1966.

At the world's fair of 1967 in Montreal the Czechoslovak pavilion showed a motion picture that allowed audiences to vote on how to proceed at five binary forks. The offer was partly a fraud because there were just two films with plots designed so that they diverged only to come together again at the next juncture. According to Theodore H. Nelson, writing on "branching movies" in *Dream Machines* (half of a two-part book that he published in Chicago in 1974), the projectionist simply dropped an opaque slide in front of whichever portion of the film was not to be seen. Without self-intersection five genuine binary choices would produce $2^5 = 32$ different films, which would be a bit expensive for filmmakers, not to mention the excessive burden on the actors.

"Homosyntaxism" is an Oulipian term for replacing all the words of a passage with new words while preserving the underlying syntax. Something similar was done by Mortimer J. Adler in *Diagrammatics*, a curious little book of nonsense prose that he and Maude P. Hutchins (then the wife of Robert Maynard Hutchins) cooked up in 1932. Adler addled the text and Mrs. Hutchins provided illustrations. I had the pleasure of hearing Adler and Mrs. Hutchins give a joint lecture (with slides) about their book in Mandel Hall at the University of Chicago. Random House published the volume in a limited edition of 750 copies, copy one going to Gene Tunney, the prizefighter. You can read all about this book and how critics reacted to it in Adler's autobiography, *Philosopher at Large* (Macmillan, 1977), pages 157–159.

It is sometimes possible to rearrange the words of a familiar passage of poetry to make an entirely different poem. Punctuation and capitalization can be varied. (A venture of this kind by Lindon and me appeared under the title "Pied Poetry" in *Word Ways*, May, 1973.) Here is one of my scramblings. Can you reconstruct the original? (The poet's name is an anagram of his real name.)

Prison Bloom and Withers?
Poison the air-well?
What good is there in that?
It is only in deeds
Vilest man
Wastes like weeds.

— S. Waldo Rice

In February 1974, *Word Ways* published a poem by Mary Young-quist, a frequent contributor, that is reproduced in Figure 45. What is so unusual about her poem?

What happens when one poet writes a poem and then gives another poet an alphabetized list of all the words in it, whereupon the second poet uses them to write a new poem without seeing the original? Oulipo members have not tried this, but *Word Ways* has published the remarkable results of several such experiments in what Eckler calls "vocabularyclept" poems. (See the issues for May 1969, August 1970, and May 1975, also Eckler's column in *Games and Puzzles*, July 1976, and Chapter 4 of Bergerson's book mentioned earlier.) It is surprising how good a new poem can be produced, with almost no resemblance to the original except in mood. Eckler has published some interesting statistical results

---

Winter Reigns

Shimmering, gleaming, glistening glow—
Winter reigns, splendiferous snow!
Won't this sight, this stainless scene,
Endlessly yield days supreme?

Eyeing ground, deep piled, delights
Skiers scaling garish heights.
Still like eagles soaring, glide
Eager racers; show-offs slide.

Ecstatic children, noses scarved—
Dancing gnomes, seem magic carved—
Doing graceful leaps. Snowballs,
Swishing globules, sail low walls.

Surely year-end's special lure
Eases sorrow we endure,
Every year renews shared dream,
Memories sweet, that timeless stream.

— Mary Youngquist

*Figure 45*
*A poem with a hidden structure*

that emerge from comparisons of the two poems. As one might expect, the shorter the original, the closer the convergence toward two identical poems.

The 11 most frequently used letters in French can be arranged to spell *ulcérations*. (It has the same meaning in French as in English.) Perec has amused himself by writing what can be called isogrammatic poems—poems written entirely with these 11 letters. He has also published poems that contain the same 11 letters plus one "free" letter. The technique is shown in Figure 46. There the position of each free letter is indicated by a cross. Following that is Perec's poem after each cross has been replaced by a letter of his choice. Mathews translates:

> Believed a bastion, the wall is far off: this smarts—you skim over shut yesterday, your nakedness, the arch where you synchronously bound the indestructible courtyard to its short, denied knell, there received. Your story: white against the night, echoes.

When the American linotype machine was invented, it was believed the 12 most commonly used letters in English, in decreasing order of frequency, are etaoin shrdlu. These two nonsense words are spelled by the first and second columns of the traditional linotype keyboard. Sometimes a printer runs a finger down the columns to make a slug, for

CRU+ASTIONLE
+URESTLOINCA
CUITONRASEL+
IERCLOSTANU+
ITELARCOUS+N
C+RONETULIAS
LACOURIN+EST
RUCTI+LEASON
+LASCOURTNIE
RECULATON+IS
TOIRE+LANCSU
RLANUITEC+OS

Cru bastion, le mur est loin: ca cuit,
on rase l'hier clos, ta nudité, l'arc
où, synchrone, tu lias la cour
indestructible à son glas court nié,
reçu là.

ton histoire:
blanc sur la nuit,
échos

**Figure 46**
*An isogram by Georges Perec*

marking purposes, that he intends to remove later. If he forgets, the cabalistic words may get printed. Perhaps readers can produce some etaoin-shrdlu poems with or without a free 13th letter.

Only one English word is known that spells with the letters of etaoin shrdlu. It is "outlandisher," to be found in the *New Standard Unabridged Dictionary*. It is also possible to use the letters to make two words that name a region of a well-known country. Can you discover the region?

The Oulipo has also worked on what in English are called "snowball sentences." Each word is one letter longer than its predecessor. Mathews gives a 22-word snowball beginning with *O le bon sens* and ending with *pseudotransfigurations*. A good English specimen, from Dmitri Borgmann's classic *Language on Vacation*, is the following 20-word snowball:

> I do not know where family doctors acquired illegibly perplexing handwriting; nevertheless, extraordinary pharmaceutical intellectuality, counterbalancing indecipherability, transcendentalizes intercommunications' incomprehensibleness.

*La Littérature Potentielle* is packed with many other kinds of linguistic play, but I have room for only a few more. Jacques Bens writes what he calls "irrational sonnets." The 14 lines are partitioned into five parts: three lines, one line, four lines, one line, five lines. These five digits, 31415, are the first five digits of pi, an irrational number—hence the name given to the sonnet. The rhyme scheme is *aab, c, baab, c, cdccd*. The two single lines must end with the same word.

Noël Arnaud writes what he calls "heterosexual poems." They are poems in which lines with a masculine ending (stressed last syllable) alternate with lines that terminate in a feminine rhyme (unstressed last syllable).

The Oulipo has developed a variety of algebraic formulas that the members apply effectively to the plot structures of novels, stories and plays. Claude Berge, an eminent graph theorist, has shown how directed graphs (graphs with arrows that give each edge a direction) can assist in analyzing literary structures.

*Les Horreurs de la Guerre (The Horrors of War)* by Perec is a three-act play in which the entire dialogue is a recital of words that give in order a single recitation of the French alphabet. *Word Ways* has published a clever analogue in English by Eckler and the following remarkable passage by Mathews. It is spoken by a crow to a scarecrow:

> Hay, be seedy! He-effigy, hate-shy jaky yellow man, oh peek, you are rusty, you've edible, you ex-wise he! (For more examples, see *Word Ways*, May 1981, page 85.)

Ingenious, I hear you say, but how frivolous, and what a sad waste of creative energy! Yet does it not bring home to us how a culture's language, with its mysterious blend of sound and meaning, is a structure with an independent life of its own? "The West is mad," writes Mathews in *The Sinking*. "I could not live without your words, no matter how you spell them." Had the crow's remark appeared somewhere in the bowels of *Finnegans Wake*, can you imagine a Joyce buff who would have found it out of place or who would not have enjoyed discovering for himself the hidden underlying form of the passage?

## ANSWERS

The scrambled poem is constructed from the following lines of Oscar Wilde's "The Ballad of Reading Gaol":

> The vilest deeds like poison weeds
> Bloom well in prison air:
> It is only what is good in Man
> That wastes and withers there.

Catherine Hoff constructed five other excellent scrambled versions of Wilde's four lines. They appeared in *Scientific American's* Letters department (April 1977), credited to poets Oswald C. Ire, Rod I. Clawse, Eric O. L. Daws, Dora S. Wilec and Rosa W. Clide. Sareen Gerson wondered why my S. Waldo Rice failed to scramble the rest of Wilde's stanza:

> Pale Anquish keeps
> The heavy gate, and the
> Warder is Despair.

She completed the stanza with:

> The gate is heavy anguish.
> Despair and warder
> The pale keeps.

The secret of Mary Youngquist's poem is that the last letter of every word is the initial letter of the next word. The property even extends to the poet's name at the end. Several readers suggested altering the poem's title so that it begins with *t*, the last letter of "Youngquist," such as "This Snow, Winter Reigns," or "Then, Now, Winter Reigns," thereby making the hidden structure cyclical. Both those titles came from Bob Gregorac.

The letters of etaoin shrdlu can be rearranged to spell South Ireland. John Rigney arranged the same letters to spell the advice given by a Spaniard about how to reach South Ireland: "Sail due North." Dolores Kozielski scrambled the letters to make the hyphenated word anti-shoulder. Richard Hawkins quoted a travel agent: "I handle tours." Tom Doyle reminded me that Etaoin Shrdlu is a character in Elmer Rice's play *The Adding Machine*.

Etaoin shrdlu is now almost obsolete as linotype machines are being replaced by computer typesetting, but QWERTY is still with us as the first six letters of the standard typewriter keyboard. By the way, did you know that TYPEWRITER spells with the top row of letters on this keyboard?

Many readers produced poems with lines formed entirely with the letters of etaoin shrdlu. One of the best, by Walter Leight, was published in the Letters department of *Scientific American* (April 1977). Titled "Sharon Dilute," it consisted of eight quatrains that begin:

> Old hunter, as I
> dash in to rule,
> insult a horde
> and so lie hurt,
>
> no adult heirs,
> riled, shout an
> oath: "Sin lured,
> had soul inert."

# CHAPTER 7

# The Oulipo II

The rose-lipt girls are sleeping;
Sleep falls; men are at peace again
In fields where roses fade
While the small drops fall softly down.

My column on the Oulipo led to an article about the group in *Time* (January 10, 1977), with a photograph of François Le Lionnais leafing though the strips of Raymond Queneau's book of $10^{14}$ sonnets. In *Time*'s Letters department (January 31), readers suggested such perverbs as "The worm is on the other foot" and "Time wounds all heels" and constructed two unusual snowball sentences.

Le Lionnais wrote to tell me about other workshop groups he had founded: Oupeinpo (painting), Oumupo (music), Oucinepo (cinema) and Oulipopo (detective literature). In 1978 his dictionary of mathematics was published and was soon followed by his anthology of remarkable numbers, *Les Nombres Remarquables* (1983), which lists and comments on some 500 unusual numbers. Plate 3 reproduces a photograph of Oulipo members in his garden, in 1975. He also reported that Claude Berge, Oulipo's graph theorist, had written a "vanishing line" sonnet modeled after the vanishing leprechaun paradox that I reproduced in Chapter 12 of my *Wheels, Life, and Other Mathematical Amusements*. When the sonnet is cut into three parts and rearranged, it changes from 15 lines to 14, making sense of course in both forms.

Georges Perec, who died from cancer in 1982 (age 46), was born in Paris of Jewish parents who left Poland in the twenties. His father lost his life in the German invasion of France, and his mother died in a concentration camp. Of his more than a dozen books, his masterpiece, *La Vie, Mode d'Emploi*, published here in 1987 as *Life: A User's Manual*, deserves some comment. Its 581 pages are broken into 99 short chapters and an epilogue that describe in detail every room in a Paris apartment building and give an account of the life of each resident. The 100 rooms correspond to the cells of a $10 \times 10$ Graeco-Latin square (for a discussion of such squares see Chapter 14 of my *New Mathematical Diversions*). We visit each room in turn along the path of a chess knight that hops from cell to cell. The rooms also are depicted in an enormous painting of their 100 interiors by an aged artist who has lived in the building for 55 years.

The French original bristles with untranslatable wordplay, mathematical recreations, chess problems and concealed quotations from famous writers. Paul Auster, reviewing the novel for the *New York Times Book Review* (November 15, 1987), called the book's traps and allusions "prodigiously entertaining," and recorded his delight in knowing by chance that when Perec refers to a melody composed by Arthur Stanley Jefferson, he is purloining the real name of comedian Stan Laurel. The book has an index of proper names, a chronology of main dates and a checklist of plots.

The 99 chapters interlock like pieces in a jigsaw puzzle, the novel's central symbol. For 10 years Percival Bartlebooth, a wealthy British resident of the Paris apartment, studies watercolor painting. During the next 20 years he travels around the world to paint 500 different seaports. Each picture is sent to another resident of the building, Gaspard Winckler, who turns each picture into a 750-piece jigsaw puzzle. After returning from his travels, Bartlebooth spends the next 20 years solving the puzzles. Each completed picture is then dipped in a solution that totally erases it. Unfortunately, while Bartlebooth is trying to solve his 439th puzzle, he dies holding a piece shaped like a W, but the one remaining hole in the puzzle has the shape of an X. Reality defeats life. As Sven Birkerts puts it in his *New Republic* review (February 8, 1988), the ending is Perec's "sly elbow-poke at his own enterprise." Perec, declared fellow Oulipoist Italo Calvino, was "one of the most singular literary personalities in the world, a writer who resembled absolutely no one else."

Calvino, himself a singular writer, was the Oulipoist who achieved the largest following in the United States. (He and Harry Mathews were both elected to the Oulipo on Valentine's Day, 1973). When he died in 1985 (age 61), he was widely regarded as Italy's most distinguished novelist. Like the fiction produced by his Oulipo friends, Calvino's novels and short stories are rich in humor, fantasy and bizarre ways of constructing plots. The three tales in *Our Ancestors* are about a knight who is split in half (one half living on as a benevolent man, the other as a cruel man); another knight who, like so many political leaders, has no existence inside his armor; and a nobleman who lives in trees. In *The Castle of Crossed Destinies* (published here in 1977), randomly played Tarot cards (their pictures are reproduced in the book's margins) provide what Calvino called "a machine for constructing stories." (See John Updike's shrewd commentary "Card Tricks" in his collection of essays *Hugging the Shore*, pages 463–470.)

Calvino's last major work, *If on a Winter's Night a Traveler* (The United States edition appeared in 1981), is about itself—a novel that the Reader (a character in the story) purchases. Unfortunately, the book he buys is defective, its pages having become mixed at the bindery with the pages of a Polish novel by another author. The bookseller exchanges the garbled volume for what is supposed to be the Polish novel, but it too is defective. The bindery has alternated printed and blank pages. This is only the beginning of a story line of increasing confusion and complexity. There are 10 different plots, each ending with cliff-hangers, or what Mary McCarthy, reviewing the novel for the *New York Review of Books* (June 25, 1981), sees as "ten regulated instances of coitus interruptus in the art and practice of fiction."

The most Oulipolian American novelist—though I don't know how much Oulipo writers have actually influenced him—is surely John Barth, with Thomas Pynchon, Robert Coover and Donald Barthelme close behind. Word and mathematical play is densest in Barth's *Letters*, but discussing this convoluted work here would take us too far afield.

A plot that forks into two or more possible endings is not a new idea—it was used effectively in plays by Lord Dunsany and J. B. Priestley and, more recently, by John Fowles at the end of his novel *The French Lieutenant's Woman*. The Oulipo carried this technique to its ultimate by allowing readers to make their own choices of alternate plot lines at various points in a narrative. Queneau's *Tale of Your Own Fashion* starts out: "Do you wish to hear the story of the three alert beans? If yes, jump to 4. If not, jump to 2." If you pick 4, the tale continues: "Once there were three little beans. . . . If you like the description, jump to 5. If you prefer another description, jump to 9." And so on.

Known as "interactive fiction," the genre became popular here in both juvenile and adult books and in a few motion pictures and stage shows. In the late seventies Bantam started a highly successful line of children's books called "Choose Your Own Adventure." In each story the reader could make choices at some two dozen points that led to a large number of possible endings. Other publishers got into the act in the eighties with similar books. Signet called its series "Lifegames for Adults." Each novel branches into 64 alternate last chapters, some happy, some terrible. Pocket Books was soon doing its "Which Way Books": Which Way Follow Your Heart Romances, Play-it-your Way Sports Books, Super Hero Which Way Books and Which Way Secret Door Books. Simon and Schuster came out with "Plot-your-own Horror Stories," and Lerner Publications introduced its "You are the Coach Books."

None of these books had any lasting merit. Nothing is easier than choosing forking paths on a computer, and so it is hardly surprising that interactive "novels" began to appear on the market as computer software. The first, *Adventure*, was a treasure hunt through dungeons and caves. Murder mysteries were naturals for this sort of thing, unraveling in thousands of different ways and sometimes taking months to solve. Science fiction became popular. Ray Bradbury helped in the writing of an interactive version of his *Fahrenheit 451*. Arthur Clarke's *Rendezvous with Rama* and Douglas Adams's *Hitchhiker's Guide to the Galaxy* were other top sellers. Michael Crichton wrote a new novel, *Amazon*, that involves a search for a lost city in the jungles of South America. Whether these computer "novels" will evolve into something that can be called literature or just be passing recreations on the comic-book level is still anybody's guess.

In the late eighties several movies and plays allowed audiences to decide how the story should continue. *The Mystery of Edwin Drood*, produced in Manhattan in 1985, permitted audiences to vote on how the play should end after the point where Dickens had left his novel unfinished.

Inspired by Oulipolian wordplay, readers sent in numerous perverbs, such as this one from Carolyn Weyant: "In one ear and gone tomorrow." Proverbs can be altered simply by switching key words around, such as "The oboe is an ill wood wind that nobody blows good." In 1969 the wife of New York City mayor John Lindsay apologized on television for her husband's seconding of Spiro Agnew's nomination for vice president. "Politics," she said, "makes strange bedfellows." Told of his wife's remark, Mayor Lindsay replied (*New York Times*, December 31): "Well, bedfellows make strange politics." "They also wait," wrote Robert Jenks on a postcard, "who only stand and serve." Proverbs altered by puns were passed along by Leonard Morgenstern: "Many are

cold but few are frozen"; "One's man Mede is another man's Persian."
From John Gummere: "Too many crooks spoil the brothel."

Al Grand wrote to add *bras* to Mathews' French-English word list.
When Grand's French classes sing "La Marseillaise," he said, the word
always produces squeals of laughter from his sixth-grade girls. The lines
*"Mugir ces féroces soldats, ils viennent jusque dans nos bras . . ."* meta-
morphose in the minds of the girls, he wrote, as something like: "When
these groaning soldiers get the 'blahs,' they come and try to snatch our
bras. . . ."

Gummere, headmaster emeritus of William Penn Charter School,
Haverford, Penn., was inspired by Mathews' French-English words to
compile a similar list of Latin words. Published in *The Classical Outlook*
(March/April 1931), it is here reproduced in Figure 47 with Gummere's
permission.

Will Shortz, writing about the National Puzzler's League in *Games*
magazine (January/February 1979), credited the following splendid ABC
sentence to a lady puzzler who uses the pseudonym Mona Lisa: "A
brilliant Chinese doctor exhorted four graduating hospital interns, 'Just
keep looking, men—no other prescription quickly relieves sore throats,
unless veterinarians willfully x-ray your zebras.'"

| | | | |
|---|---|---|---|
| a | do[2] | is | post |
| acre | dole | it | quid[4] |
| age | dote | late | re[2] |
| ages | ducat | male | rear |
| ago | eat | mallet | rue |
| an | era | mane | sere |
| at | ere | mare | sex |
| boa | fare | mi[1] | si[1] |
| bone | ferret | miles | sic[5] |
| cadet | fit | mire | sol[1] |
| cane | flare | more | sole[6] |
| cave | flat | net | stare |
| clam | for | nix | sue |
| cur | fore | no | sum |
| dare | fur | pace | tale |
| date | graves | pane | tam |
| dens | hem | pellet | time |
| die | his | pone[3] | tot |
| do[1] | I | possum | violet |

(NOTES: 1, musical note; 2, to act; 3, corn pone;
4, a chew; 5, urge to attack; 6, the fish)

**Figure 47**  *John Gummere's Latin-English list*

Albert L. Ely, Jr., interwove the quatrains by Elinor Wylie and Edna St. Vincent Millay with A. E. Housman's "A Shropshire Lad" to obtain the following stanzas:

> With rue my heart is laden —
> My candle burns at both ends
> For golden friends I had.
> It will not last the night
> For many a rose-lipt maiden,
> But ah, my foes, and oh, my friends,
> And many a lightfoot lad,
> It gives a lovely light.
>
> By brooks too broad for leaping
> Sleep falls with limpid drops of rain.
> The lightfoot boys are laid
> Upon the steep cliffs of the town;
> The rose-lipt girls are sleeping;
> Sleep falls; men are at peace again
> In fields where roses fade
> While the small drops fall softly down.

Dick Ringler, inspired by J. A. Lindon's palindromic dialogue between Adam and Eve, composed the following pair of quatrains, each a word reversal of the other:

DIPTYCH

To Theseus: Finding No Minotaur

> Thread the chaos, pattern the despair.
>   Shadows loom and worry you:
> Dead hope, and empty heaven, and now bare
>   Meadows — fearful! but all perspective true.

To Penelope: Weaving in Autumn

> True perspective all, but fearful! meadows
>   Bare now, and heaven empty, and hope dead,
> You worry and loom shadows,
>   Despair the pattern, chaos the thread.

A correspondent who wishes to remain nameless asked if I was aware that Adam and Eve were not Jewish, but Irish. It seems that when they first met, each lifted up the other's fig leaf. "O'Hare!" exclaimed Adam, while Eve shouted "O'Toole!"

*Country Cooking and Other Stories*, a collection of short tales by Harry Mathews, was published in 1980 by Burning Deck, a firm in Providence, R.I. *The Review of Contemporary Fiction* devoted its Fall 1987 issue to Mathews (see bibliography), and that same year, November 4–6, a literary colloquium on the Oulipo was sponsored by the Romance languages department of Baruch College in New York City. There were readings from Oulipo authors and lectures about and discussions of several recent Oulipo works, including Mathews' 1987 novel *Cigarettes*. This amusing, intricately plotted novel is unusual in lacking the concealed structures, subtle puzzles and outrageous wordplay of Mathews' earlier fiction and poetry. Robert Towers, reviewing *Cigarettes* in the *New York Review of Books* (January 21, 1988), called it "cool and elegant" and an "odd and gratifying novel."

## BIBLIOGRAPHY

*Oddities and Curiosities of Words and Literature*. C. C. Bombaugh, Martin Gardner, ed. Dover, 1961.

*Language on Vacation: An Olio of Orthographical Oddities*. Dmitri Borgmann. Scribner's, 1965.

*Beyond Language: Adventures in Word and Thought*. Dmitri Borgmann. Scribner's, 1967.

*The Game of Words*. Willard Espy. Grosset and Dunlap, 1972.

*La Littérature Potentielle*. Oulipo, Gallimard, 1973.

*Palindromes and Anagrams*. Howard Bergerson. Dover, 1973.

*Almanac of Words and Play*. Willard Espy. Potter, 1975.

"Oulipo." Harry Mathews, in *Word Ways*, 9, May 1976, pp. 67–74.

"Oulipo." Ross Eckler, in *Word Recreations*. Dover, 1979, pp. 4–9.

*Another Almanac of Words and Play*. Willard Espy. Potter, 1980.

*Atlas de Littérature Potentielle*. Oulipo. Gallimard, 1981.

*The Oxford Guide to Word Games*. Tony Augarde. Oxford, 1984.

"Puzzles in Ulysses." Martin Gardner, in *Semiotica*, 57, 1985, pp. 317–330.

*Names*. Paul Dickson. Delacorte, 1986.

*Names and Games*. Ross Eckler, ed. University Press of America, 1986.

*Harry Mathews Number. The Review of Contemporary Fiction*, 7, Fall 1987. The issue is devoted to tributes to Mathews by 18 writers and includes a lengthy checklist of Mathews's writings and articles about him.

"That Ephemeral Thing." A review of Georges Perec's *Life: A User's Manual*, by Harry Mathews, in the *New York Review of Books*, June 16, 1988, pp. 34–37.

*Word Recreations*. Ross Eckler. St. Martin's Press, 1996.

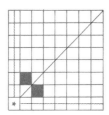

# CHAPTER 8

# Wythoff's Nim

An analysis of a simple two-person game can lead into fascinating corners of number theory. We begin with a charming, little-known game played on a chessboard with a single queen. Before we are through, we shall have examined a remarkable pair of number sequences that are intimately connected with the golden ratio and generalized Fibonacci sequences.

The game, which has no traditional name, was invented about 1960 by Rufus P. Isaacs, a mathematician at Johns Hopkins University. It is described briefly (without reference to chess) in Chapter 6 of the 1962 English translation of *The Theory of Graphs and Its Applications*, a book in French by Claude Berge. (We met Berge in the previous chapter as a member of the Oulipo.) Let's call the game "Corner the Lady."

Player *A* puts the queen on any cell in the top row or in the column farthest to the right of the board; the cells appear in gray in Figure 48. The queen moves in the usual way but only west, south or southwest. Player *B* moves first, then the players alternate moves. The player who gets the queen to the starred cell at the lower left corner is the winner.

No draw is possible, so that *A* or *B* is sure to win if both sides play rationally. It is easy to program an HP-97 printing calculator or the HP-67 pocket calculator to play a perfect game. Indeed, a magnetic card

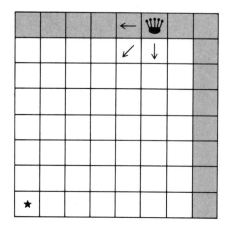

**Figure 48**
*The cornering game of Rufus P. Isaacs*

supplied with Hewlett-Packard's book *HP-67/HP-97 Games Pac 1* provides just such a program.

Isaacs constructed a winning strategy for cornering the queen on boards of unbounded size by starting at the starred cell and working backward. If the queen is in the row, column or diagonal containing the star, the person who has the move can win at once. Mark these cells with three straight lines as is shown in part A of Figure 49. It is clear that the two shaded cells are "safe," in the sense that if you occupy either one, your opponent is forced to move to a cell that enables you to win on the next move.

Part B of the illustration shows the next step of our recursive analysis. Add six more lines to mark all the rows, columns and diagonals containing the two previously discovered safe cells. This procedure allows us to shade two more safe cells as shown. If you occupy either one, your opponent is forced to move, so that on your next move you can either win at once or move to the pair of safe cells nearer the star.

Repeating this procedure, as is shown in part C of the illustration, completes the analysis of the chessboard by finding a third pair of safe cells. It is now clear that Player *A* can always win by placing the queen on the shaded cell in either the top row or the column farthest to the right. His strategy thereafter is simply to move to a safe cell, which he can always do. If *A* fails to place the queen on a safe cell, *B* can always win by the same strategy. Note that winning moves are not necessarily unique. There are times when the player with the win has two choices; one may delay the win, the other may hasten it.

Our recursive analysis extends to rectangular matrixes of any size or shape. In part D of the illustration, a square with 25 squares on a side is

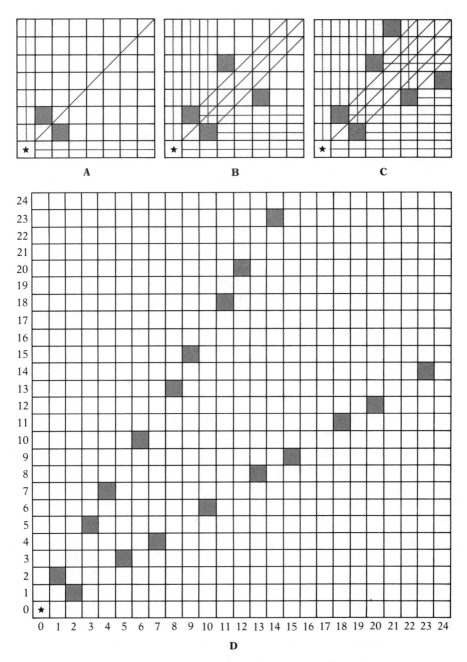

***Figure 49*** *(A, B, C) A recursive analysis of "Corner the Lady" (D) The first nine pairs of safe cells*

shown with all the safe cells shaded. Note that they are paired symmetrically with respect to the main diagonal and lie almost on two lines that fan outward to infinity. Their locations along those lines seem to be curiously irregular. Are there formulas by which we can calculate their positions nonrecursively?

Before answering let us turn to an old counter take-away game said to have been played in China under the name *tsyan-shidzi*, which means "choosing stones." The game was reinvented by the Dutch mathematician W. A. Wythoff, who published an analysis of it in 1907. In Western mathematics it is known as "Wythoff's Nim."

The game is played with two piles of counters, each pile containing an arbitrary number of counters. As in Nim, a move consists in taking any number of counters from either pile. At least one counter must be taken. If a player wishes, he may remove an entire pile. A player may take from both piles (which he may not in Nim), provided that he takes the same number of counters from each pile. The player who takes the last counter wins. If both piles have the same number of counters, the next player wins at once by taking both piles. For that reason the game is trivial if it starts with equal piles.

We are ready for our first surprise. Wythoff's Nim is isomorphic with the Queen-Cornering game! When Isaacs invented the game, he did not know about Wythoff's Nim, and he was amazed to learn later that his game had been solved as early as 1907. The isomorphism is easy to see. As is shown in part D of Figure 49, we number the 25 columns along the $x$ coordinate axis, starting with 0; the rows along the $y$ coordinate axis are numbered the same way. Each cell can now be given an $x/y$ number. These numbers correspond to the number of counters in piles $x$ and $y$. When the queen moves west, pile $x$ is diminished. When the queen moves south, pile $y$ is diminished. When it moves diagonally southwest, both piles are diminished by the same amount. Moving the queen to cell 0/0 is equivalent to reducing both piles to 0.

The strategy of winning Wythoff's Nim is to reduce the piles to a number pair that corresponds to the number pair of a safe cell in the Queen game. If the starting pile numbers are safe, the first player loses. He is certain to leave an unsafe pair of piles, which his opponent can always reduce to a safe pair on his next move. If the game begins with unsafe numbers, the first player can always win by reducing the piles to a safe pair and continuing to play to safe pairs.

The order of the two numbers in a safe pair is not important. This condition corresponds to the symmetry of any two cells on the chessboard with respect to the main diagonal: they have the same coordinate numbers, one pair being the reverse order of the other. Let us take the safe pairs in sequence, starting with the pair nearest 0/0, and arrange them in a row with each smaller number above its partner, as in Figure 50. Above the pairs write their "position numbers." The top numbers of the safe pairs form a sequence we shall call A. The bottom numbers form a sequence we shall call B.

These two sequences, each one strictly increasing, have so many remarkable properties that dozens of technical papers have been written about them. Note that each B number is the sum of its A number and its position number. If we add an A number to its B number, the sum is an A number that appears in the A sequence at a position number equal to B. (An example is $8 + 13 = 21$. The 13th number of the A sequence is 21.)

We have seen how the two sequences are obtained geometrically by drawing lines on the chessboard and shading cells according to a recursive algorithm. Can we generate the sequences by a recursive algorithm that is purely numerical?

We can. Start with 1 as the top number of the first safe pair. Add this to its position number to obtain 2 as the bottom number. The top number of the next pair is the smallest positive integer not previously used. It is 3. Below it goes 5, the sum of 3 and its position number. For the top of the third pair write again the smallest positive integer not yet used. It is 4. Below it goes 7, the sum of 4 and 3. Continuing in this way will generate series A and B.

There is a bonus. We have discovered one of the most unusual properties of the safe pairs. It is obvious from our procedure that every positive integer must appear once and only once somewhere in the two sequences.

Is there a way to generate the two sequences nonrecursively? Yes. Wythoff was the first to discover that the numbers in sequence A are simply multiples of the golden ratio rounded down to integers! (He wrote that he pulled this discovery "out of a hat.")

The golden ratio, as most readers of this book are aware, is one of the most famous of all irrational numbers. Like pi it has a way of appearing in unlikely places. Ancient Greek mathematicians called it the "extreme and mean ratio" for the following reason. Divide a line segment into

| Position ($n$) | 1 | 2 | 3 | 4 | 5 | 6 | 7 | 8 | 9 | 10 | 11 | 12 | 13 | 14 | 15 |
|---|---|---|---|---|---|---|---|---|---|---|---|---|---|---|---|
| A. $[n\phi]$ | 1 | 3 | 4 | 6 | 8 | 9 | 11 | 12 | 14 | 16 | 17 | 19 | 21 | 22 | 24 |
| B. $[n\phi^2]$ | 2 | 5 | 7 | 10 | 13 | 15 | 18 | 20 | 23 | 26 | 28 | 31 | 34 | 36 | 39 |

**Figure 50** *The first 15 safe pairs in Wythoff's Nim*

parts $A$ and $B$ in such a way that the ratio of length $A$ to length $B$ is the same as the ratio of the entire line to $A$. You have divided the line into a golden ratio. Because this has been widely thought to be the most pleasing way to divide a line, the golden ratio has provoked a bulky literature (much of it crankish) about the use of the ratio in art and architecture.

We can calculate the golden ratio by assigning a length of 1 to line segment $B$. Our method of dividing the line is expressed by $(A + 1)/A = A/1$, a simple quadratic equation that produces for $A$ a positive value of $(1 + \sqrt{5})/2 = 1.61803398 \ldots$, the golden ratio. Its reciprocal is $0.61803398. \ldots$ It is the only positive number that becomes its own reciprocal when 1 is taken from it and that becomes its own square when 1 is added to it. Its negative reciprocal has the same properties. In Britain the golden ratio is usually signified by the Greek letter $\tau$ (tau). I shall follow the American practice of calling it $\phi$ (phi).

The numbers in sequence $A$ are given by the formula $[n\phi]$, where $n$ is the position number and the brackets signify discarding the fractional part. $B$ numbers can be obtained by adding $A$ numbers to their position numbers, but it turns out that they are rounded-down multiples of the square of phi. The formula for sequence $B$, therefore, is $[n\phi^2]$. The fact that every positive integer appears once and only once among the safe pairs can be expressed by the following remarkable theorem: The set of integers that lie between successive multiples of phi and between successive multiples of phi squared is precisely the set of natural numbers.

Two sequences of increasing positive integers that together contain every positive integer just once are called "complementary." Phi is not the only irrational number that generates such sequences, although it is the only one that gives the safe pairs of Wythoff's Nim. In 1926 Sam Beatty, a Canadian mathematician, published his astounding discovery that any positive irrational number generates complementary sequences.

Let $k$ be the irrational number, with $k$ greater than 1. Sequence $A$ consists of multiples of $k$, rounded down, or $[nk]$, where $n$ is the position number and the brackets indicate discarding the fraction. Sequence $B$ consists of rounded-down multiples of $k/(k - 1)$, or $[nk/(k - 1)]$. Complementary sequences produced in this way are called Beatty sequences. If $k$ is phi, the second formula gives rounded-down multiples of $1.618 + /0.618 + = 2.618+$, which, owing to the whimsical nature of phi, is the square of phi. Readers might like to convince themselves that Beatty's formulas do indeed produce complementary sequences by letting $k = \sqrt{2}$, pi, $e$ or any other irrational, and that rational values for $k$ fail to produce such sequences.

Whenever the golden ratio appears, it is a good bet that Fibonacci numbers lurk nearby. The Fibonacci sequence is 1, 1, 2, 3, 5, 8, 13, 21, 34 . . . , in which each number after the first two is the sum of the two preceding numbers. A general Fibonacci sequence is defined in the same way, except that it can begin with any pair of numbers. A property of every Fibonacci sequence of positive integers is that the ratio of adjacent terms gets closer and closer to phi, approaching the golden ratio as a limit.

If we partition the primary Fibonacci sequence into pairs, 1/2, 3/5, 8/13, 21/34 . . . , it can be shown that every Fibonacci pair is a safe pair in Wythoff's Nim. The first such pair not in this sequence is 4/7. If we start another Fibonacci sequence with 4/7, however, and partition it 4/7, 11/18, 29/47 . . . , all these pairs are also safe in Wythoff's Nim. Indeed, these pairs belong to a Fibonacci sequence of what are called Lucas numbers that begins 2, 1, 3, 4, 7, 11. . . .

Imagine that we go through the infinite sequence of safe pairs (in the manner of Eratosthenes' sieve for sifting out primes) and cross out the infinite set of all safe pairs that are pairs in the Fibonacci sequence. The smallest pair that is not crossed out is 4/7. We can now cross out a second infinite set of safe pairs, starting with 4/7, that are pairs in the Lucas sequence. An infinite number of safe pairs, of which the lowest is now 6/10, remain. This pair too begins another infinite Fibonacci sequence, all of whose pairs are safe. The process continues forever. Robert Silber, a mathematician at North Carolina State University, calls a safe pair "primitive" if it is the first safe pair that generates a Fibonacci sequence. He proves that there are an infinite number of primitive safe pairs. Since every positive integer appears exactly once among the safe pairs, Silber concludes that there is an infinite sequence of Fibonacci sequences that exactly covers the set of natural numbers.

Take the primitive pairs 1/2, 4/7, 6/10, 9/15 . . . in order and write down their position numbers, 1, 3, 4, 6. . . . Does this sequence look familiar? As Silber shows, it is none other than sequence A. In other words, a safe pair is primitive if and only if its position number is a number in sequence A.

Suppose you are playing Wythoff's game with a very large number of counters or on a chessboard of enormous size. What is the best way to determine whether a position is safe or unsafe, and how do you play perfectly if you have the win?

You can, of course, use the phi formulas to write out a sufficiently large chart of safe pairs, but this is hard to do without a calculator. Is there a simpler way comparable to the technique of playing perfect Nim by writing the pile numbers in binary notation? Yes, there is, but it uses a more eccentric type of number representation called Fibonacci notation that has been intensively studied by Silber and his colleague Ralph Gellar and also by other mathematicians such as Leonard Carlitz of Duke University.

Write the Fibonacci sequence from right to left as is shown in Figure 51. Above it number the positions from right to left. With the aid of this chart we can express any positive integer in a unique way as the sum of Fibonacci numbers. Suppose we want to write 17 in Fibonacci notation. Find the largest Fibonacci number that is not greater than 17 (it is 13) and put a 1 below it. When we move to the right, we find the next number that, added to 13, gives a sum that does not exceed 17. It is 3, and so a 1 goes below 3. When we move to the right again, the next number that gets a 1 is the 1 in the second position. The unused Fibonacci numbers get 0's.

The result is 1001010, a unique representation of 17. To translate it back to decimal notation sum the Fibonacci numbers indicated by the positions of the 1's: $13 + 3 + 1 = 17$. The 1 farthest to the right in the Fibonacci sequence is never used, so that all numbers in Fibonacci notation end in 0. It is also easy to see there are never two adjacent 1's. If there were, they would have a sum equal to the next Fibonacci number on the left, and our rules would give that number a 1 and give 0's to the original pair of adjacent 1's.

In Fibonacci notation the sum of a safe pair is the $B$ number with 0 appended. From this it follows that the Fibonacci sequence is obtained by starting with 10 and adding 0's: 10, 100, 1000, 10000. . . . The same procedure gives any Fibonacci sequence generated by a primitive pair. For example, the Lucas sequence starting with 4/7 is 1010, 10100, 101000, 1010000. . . .

Every $A$ number in Fibonacci notation has the 1 farthest to the right at an even position from the right. Every $B$ number is obtained by adding 0 to the right of its $A$ partner. Therefore every $B$ number has the 1 farthest to the right in an odd position. Since every counting number is either an $A$ number or a $B$ number, we have a simple way of deciding

| | 10 | 9 | 8 | 7 | 6 | 5 | 4 | 3 | 2 | 1 | | |
|---|---|---|---|---|---|---|---|---|---|---|---|---|
| . . . | 55 | 34 | 21 | 13 | 8 | 5 | 3 | 2 | 1 | 1 | | |
| | | | | 1 | 0 | 0 | 1 | 0 | 1 | 0 | = | 17 |

**Figure 51**  *Fibonacci notation for 17*

whether a given position in Wythoff's Nim is safe or unsafe. Write the two numbers in Fibonacci notation. If the smaller one is an $A$ number, and if adding 0 produces the other number, the position is safe; otherwise it is unsafe.

An example of the method is $8/13 = 100000/1000000$. The 1 in 100000 is at position 6, an even position, so that 100000 is an $A$ number. Adding 0 produces $1000000 = 13$, the partner of 8. We know that $8/13$ is safe. If it is your turn, your opponent has the win. If you think he cannot play perfectly, make a small random move and hope that he soon will make a mistake.

If the pair is unsafe and it is your turn, how can you determine the safe position to which you must play? There are three cases to consider. In each case call the unsafe pair $x/y$, with $x$ the smaller number, and write both numbers in Fibonacci notation.

In the first case $x$ is a $B$ number. Move to reduce $y$ to the number equal to the number obtained by deleting the right-hand digit of $x$. For example, $x/y = 10/15 = 100100/1000100$. Since 100100 has the 1 farthest to the right at an odd position, it is a $B$ number. Delete its last digit to obtain $10010 = 6$. The safe numbers you must produce (by removing from the larger pile) are 10 and 6. On a chessboard this corresponds to an orthogonal queen move.

In the second case $x$ is an $A$ number, but $y$ exceeds the number obtained by appending 0 to $x$. Move to reduce the value of $y$ to that number, for example, $x/y = 9/20 = 100010/1010100$. Because $x$'s 1 farthest to the right is in an even position, it is an $A$ number. Appending 0 produces $1000100 = 15$. This is less than 20. Therefore the safe pair to play to is $9/15$. On the chessboard this too is an orthogonal queen move.

If the numbers do not conform to cases 1 and 2, do the following:

1. Find the positive difference between $x$ and $y$.
2. Subtract 1, express the result in Fibonacci notation and change the last digit to 1.
3. Append 0 to get one number. Append two 0's to get a second number. These two numbers are the safe pair you seek, even though the resulting Fibonacci numbers may be "noncanonical" in having consecutive 1's.

An example of the third case is $x/y = 24/32 = 10001000/10101000$. The first and second cases do not apply. The difference between 24 and 32 is 8. Subtracting 1 leaves 7. In Fibonacci notation 7 is 10100. Changing the last digit to 1 produces 10101. Appending 0 and 00 yields the safe pair 101010/1010100 = 12/20. This result is reached by taking 12 from both piles. It corresponds to a diagonal queen move.

It is impossible to go into the whys of Silber's bizarre strategy. Interested readers will find the proofs in Silber's 1977 paper, "Wythoff's Nim and Fibonacci Representations." Neither can I go into the ways in which Wythoff's game has been generalized, but a word or two should be added about the game's reverse, or misère, form: the last person to play loses. As T. H. O'Beirne makes clear in *Puzzles and Paradoxes*, misère Wythoff's Nim, like misère Nim, requires only a trivial alteration of the chart of safe pairs. Remove the first pair, 1/2, and substitute 0/1 and 2/2. The misère strategy is exactly like the standard strategy except that at the end you may have to play to 2/2 or 0/1 instead of 1/2.

Let us modify Wythoff's Nim as follows. A player may take any positive number of counters from either pile, or he may take one counter from one pile and two counters from the other. Can the reader determine the chessboard model and the winning strategy?

## ANSWERS

The task was to analyze a game (similar to Nim) in which players may take from either of two piles or take one counter from one pile and two counters from the other. The last person to play wins. In the unbounded-chessboard model explained in the chapter, the first rule is equivalent to the move of a rook west or south and the second rule is equivalent to a knight jumping southwest. The take-away game is therefore isomorphic with the game of cornering a chess piece that combines the powers of rook and knight. Among enthusiasts of unorthodox, or "fairy," chess such a piece is sometimes called a "chancelor" or sometimes an "empress."

If the piece moves only like a rook, the game on the chessboard is the same as standard Nim with two piles. Safe pairs are any two equal positive integers. They correspond to cells on the board's main diagonal that passes through corner cells 0/0 and 7/7. The player who places the rook (on the top row or the column farthest to the right) wins only by putting it on 7/7. Thereafter his strategy is always to move to the diagonal. In the take-away game this means keeping the piles equal. The safe pairs are simply 1/1, 2/2, 3/3. . . .

Surprisingly, giving the rook the additional power of a knight has no effect on this strategy. Applying the recursive technique explained earlier, we find that the safe cells (or safe pairs) are exactly the same as in the rook game.

The misère form of rook-knight Nim (the last person to play loses) is more interesting. The safe pairs are 0/1, 2/3, 4/5, 6/7. . . . On a chessboard these unordered pairs are the cells shown in gray in Figure 52. The "placer" has the win, but he must put the rook-knight on a cell adjacent to the cell in the top right corner. Thereafter he moves to occupy a safe cell. This procedure eventually brings him to 0/1 or 1/0, forcing his opponent to make the final move.

Readers might enjoy analyzing the game on a standard chessboard when the placed piece has other chess powers, in each case limiting moves to west, south and southwest. A "superqueen" or "amazon" (combining queen and knight) means a loss for the placer in standard and reverse play. A king loses for the placer in standard play but wins in misère. The same result emerges if the piece is a king-knight or a king-rook. The placer wins in both types of play if the piece is a king-bishop.

## ADDENDUM

Figure 53 shows in gray the safe cells for the king, rook and bishop Nim. The bishop game is trivialized by the fact that the bishop cannot legally move to the target from any square off the main diagonal. If we restrict the bishop to this diagonal, the second player obviously wins the standard game and loses the reverse game.

We can ignore combining the powers of queen and king, queen-bishop or queen-rook or combining rook and bishop, because such

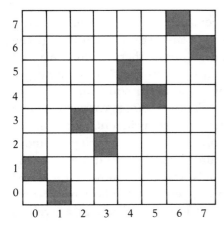

**Figure 52**
*Safe cells of reverse rook-knight Nim*

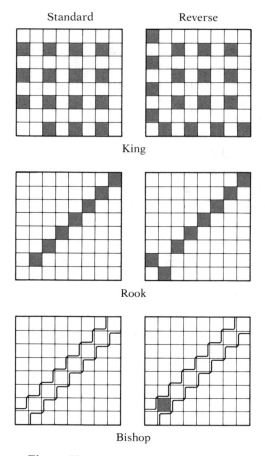

**Figure 53** *King, rook and bishop Nim*

pieces are clearly equivalent to the queen. Safe cells are shown in Figure 54 for the superqueen or amazon (queen-knight), superking (king-knight), king-rook (in *shogi*, or Japanese chess, there is such a piece called the *rya-ou*) and the king-bishop (*ryu-ma* in *shogi*). In all cases we assume that a piece can move only west, south or southwest.

Combining bishop and knight produces a piece known to some fairy-chess buffs as the "abbot," to others as the "princess." Christopher Arata sent a detailed analysis of Nimlike games to be played, under various rules, with this piece on an unlimited board. If we limit the playing field to those cells from which it is possible to move to the target square, the top of Figure 55 shows the safe cells for standard and reverse play. Arata suggested allowing the abbot to move southeast as well as

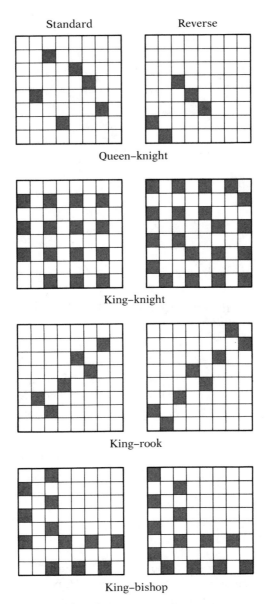

**Figure 54** *Pieces with combined powers*

southwest, in which case the safe cells become those shown at the bottom of the illustration. It is not known how to state rules that generalize these patterns to unlimited boards.

All these games have, of course, corresponding rules for playing Nim

Standard                    Reverse

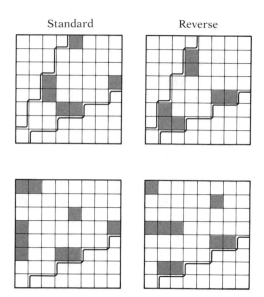

**Figure 55** *The abbot (bishop-knight) with S, W and SW moves (top), and SE added* (bottom)

with two piles of counters. For other ways of modifying Wythoff's game, readers are referred to papers listed in the bibliography. Many readers pointed out ways in which Silber's algorithm for calculating the winning strategy in Wythoff's Nim can be simplified for efficient computer programs.

## BIBLIOGRAPHY

*On Wythoff's game*

"The Golden Section, Phyllotaxis, and Wythoff's Game." H. S. M. Coxeter, in *Scripta Mathematica*, 19, June/September 1953, pp. 135–143.

*The Theory of Graphs and Its Applications*, Chapter 6. Claude Berge. Paris: Dunod, 1958; London: Methuen, 1962.

"A Generalization of Wythoff's Game." I. G. Connell, in *The Canadian Mathematics Bulletin*, 2, 1959, pp. 181–190.

"Fibonacci Nim." Michael Whiniham, in *The Fibonacci Quarterly*, 1, December 1963, pp. 9–14.

*Mathematical Games and Pastimes*. A. P. Domoryad. Macmillan, 1964, pp. 62–65.

*Puzzles and Paradoxes.* T. H. O'Beirne. Oxford University Press, 1965, pp. 131–138. Reprinted by Dover, 1984.

*Challenging Mathematical Problems with Elementary Solutions,* A. M. Yaglom and I. M. Yaglom. Vol. 2, Holden-Day, 1967, pp. 20 and 105.

"Some generalizations of Wythoff's Game and Other Related Games." John Holladay, in *Mathematics Magazine,* 41, January 1968, pp. 7–13.

"Partitions of $N$ into Distinct Fibonacci Numbers." David Klarner, in *The Fibonacci Quarterly,* 6, February 1968, pp. 235–244.

"A Generalization of Wythoff's Game." A. S. Fraenkel and I. Borash, in *Journal of Combinatorial Theory,* Series A, September 1973, pp. 175–191.

"A Fibonacci Property of Wythoff Pairs." Robert Silber, in *The Fibonacci Quarterly,* 14, November 1976, pp. 380–384.

"Wythoff's Nim and Fibonacci Representations." Robert Silber, in *The Fibonacci Quarterly,* 15, February 1977, pp. 85–88.

"Wythoff's Pairs." A. F. Horadam, in *The Fibonacci Quarterly,* 16, April 1978, pp. 147–151.

"A Property of Wythoff Pairs." V. E. Hoggatt, Jr., and A. P. Hillman, in *The Fibonacci Quarterly,* 16, October 1978, p. 472.

*Mathematical Recreations and Essays.* W. W. Rouse Ball and H. S. M. Coxeter. Dover, 1987, pp. 39–40.

"A Generalization of Wythoff's Game." V. E. Hoggatt, Jr., Marjorie Bicknell-Johnson and Richard Sarsfield, in *The Fibonacci Quarterly,* 17, October 1979, pp. 198–211.

"Some Extensions of Wythoff Pair Sequences." G. E. Bergum and V. E. Hoggatt, Jr., in *The Fibonacci Quarterly,* 18, February 1980, pp. 28–32.

"How to Beat Your Wythoff Games' Opponent on Three Fronts." Aviezri Fraenkel, in the *The American Mathematical Monthly,* 89, June/July 1982, pp. 353–361.

"Wythoff Games, Continued Fractions, Cedar Trees and Fibonacci Sequences." Aviezri Fraenkel, in *Theoretical Computer Science,* 29, 1984, pp. 49–73.

"Generalized Wythoff Numbers from Simultaneous Fibonacci Representations." Marjorie Bicknell-Johnson, in the *Fibonacci Quarterly,* 23, November 1985, pp. 308–318.

"The Sprague-Grundy Function for Wythoff's Game." Uri Blass and Aviezri Fraenkel, in *Theoretical Computer Science.* Elsevier, 1990.

### On Beatty sequences

"Problem 3173." Sam Beatty, in *The American Mathematical Monthly,* 33, 1926, p. 159; solutions in 34, 1927, p. 159.

"On a Problem Arising Out of the Theory of a Certain Game." J. V. Uspensky, in *The American Mathematical Monthly,* 34, 1927, pp. 516–521.

"Inverse and Complementary Sequences of Natural Numbers." J. Lambek and Leo Moser, in *The American Mathematical Monthly,* 61, 1954, pp. 454–458.

"On a Theorem of Uspensky." R. L. Graham, in *The American Mathematical Monthly,* 70, April 1963, pp. 407–409.

"Functions Which Represent All Integers." E. N. Gilbert, in *The American Mathematical Monthly,* 70, September 1963, pp. 736–738.

"The Bracket Function and Complementary Sets of Integers." Aviezri Fraenkel, in *The Canadian Journal of Mathematics,* 21, 1969, pp. 6–27.

"Complementary Sequences." Ross Honsberger, in *Ingenuity in Mathematics,* Chapter 7. Mathematical Association of America, 1970.

"Characterization of the Set of Values $f(n) = [n\alpha]$, $n = 1, 2, \ldots$" Aviezri Fraenkel, Jonathan Levitt and Michael Shimsoni, in *Discrete Mathematics,* 2, 1972, pp. 335–345.

"Complementary and Exactly Covering Sequences." Aviezri Fraenkel, in *Journal of Combinatorial Theory* 14, January 1973, pp. 8–20.

"A Characteristic of Exactly Covering Congruences." Aviezri Fraenkel, in *Discrete Mathematics,* 4, 1973, pp. 359–366.

"Further Characterizations and Properties of Exactly Covering Congruences." Aviezri Fraenkel, in *Discrete Mathematics,* 12, 1975, pp. 93–100.

"Beatty Sequences, Continued Fractions, and Certain Shift Operators." Kenneth Stolarsky, in *The Canadian Mathematical Bulletin,* 19, 1976, pp. 473–482.

"Complementary Systems of Integers." Aviezri Fraenkel, in *The American Mathematical Monthly,* 84, February 1977, pp. 114–115.

"A Set of Generalized Fibonacci Sequences Such That Each Natural Number Belongs to Exactly One." Kenneth Stolarsky, in *The Fibonacci Quarterly,* 15, October 1977, p. 224.

"Spectra of Numbers." R. L. Graham, Shen Lin, and Chio-Shih Lin, in *Mathematics Magazine,* 51, May 1978, pp. 174–176.

"Nonhomogeneous Spectra of Numbers." M. Boshernitzan and A. S. Fraenkel, in *Discrete Mathematics,* 34, 1981, pp. 325–327.

"A Linear Algorithm for Nonhomogeneous Spectra of Numbers." M. Boshernitzan and A. S. Fraenkel, in *Journal of Algorithms,* 5, 1984, pp. 187–198.

"Disjoint Covering Systems of Rational Beatty Sequences." Marc A. Berger, Alexander Felzenbaum and Aviezri Fraenkel, in *Journal of Combinatorial Theory,* Series A, 42, May 1986, pp. 150–153.

# 9

# Pool-Ball Triangles and Other Problems

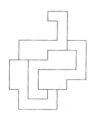

## 1. POOL-BALL TRIANGLES

Colonel George Sicherman of Buffalo tells me that about 10 years ago, while he was watching a game of pool, the following problem occurred to him: Is it possible to form a "difference triangle" in arranging the 15 balls in the usual triangular configuration before the start of a game? In a difference triangle the numbers 1 through 15 are arranged in such a way that each number below a pair of numbers is the positive difference between that pair.

It is evident from Figure 56 that the problem is trivial with the balls numbered 1, 2 and 3 and that two solutions can be obtained with them. The illustration also shows the four solutions for six balls and the four for ten balls. To Sicherman's surprise the 15 pool balls have only one basic solution. (It can, of course, be reflected.) Can you find it?

Searching for the solution is considerably simplified by first exploring triangular patterns of even and odd to see which patterns have exactly eight odd and seven even spots. It does not take long to discover that there are only five arrangements for the triangle's top row: EEOEO,

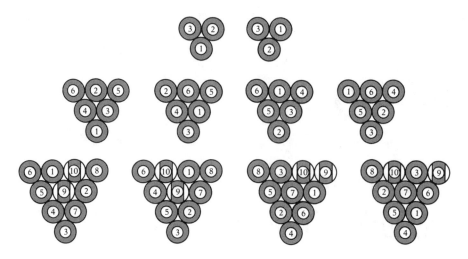

**Figure 56** *Difference triangles for three, six and ten pool balls*

OEEEO, OOOEE, OOOOE and OOEEO. The 15 ball obviously must be in the top row and the 14 ball must be in the same row or below 15 and 1. Other dodges shorten an exhaustive analysis.

The problem is related to one posed by Hugo Steinhaus in his book *One Hundred Problems in Elementary Mathematics* (Dover, 1979, a translation from an earlier Polish edition). Given a triangular array with an even number of spots, is it always possible to form an even-odd difference pattern in which the number of even spots equals the number of odd ones? The problem remained unsolved for more than a decade until Heiko Harborth, in the *Journal of Combinatorial Theory (A)*, 12, 1972, pages 253–259, proved that the answer is yes.

As far as I know, no work has been done on what we might call the general pool-ball problem. Given any triangular number of balls, numbered consecutively from 1, is it always possible to form a difference triangle? If not, is there a largest triangle for which a solution is possible? If there is, what is it? We now know that odd-even patterns for such solutions exist for all triangles with an even number of balls. Do they also exist for all triangles with an odd number of balls?

Let me add the following joke problem for readers who succeed in solving the 15-ball problem. Suppose the balls bear the 15 consecutive even numbers from 2 through 30. Is it possible to arrange the set in a difference triangle?

## 2. TOROIDAL CANNIBALISM

A torus is a surface shaped like a doughnut. Imagine a torus made of sheet rubber. It is well known that if there is a hole in such a torus, the torus can be turned inside out through the hole.

John Stillwell, a mathematician at Monash University in Australia, poses the following problem. Two toruses, *A* and *B*, are linked as is shown in Figure 57. There is a "mouth" (a hole) in *B*. We can stretch, compress and deform either torus as radically as we please, but of course no tearing is allowed. Can *B* swallow *A*? At the finish *B* must have its original shape, although it will be larger, and *A* must be entirely inside it.

## 3. EXPLORING TETRADS

The most sensational news in recreational mathematics in 1976 was surely the announcement by two University of Illinois mathematicians that they had proved the four-color-map conjecture. This famous conjecture is often confused with a simpler theorem in topology, which is easily proved; it states that no more than four regions on the plane can have a mutual border. Michael R. W. Buckley, in the *Journal of Recreational Mathematics*, 8 (1975), proposed the name tetrad for four simply connected planar regions, each pair of which shares a finite portion of a common boundary.

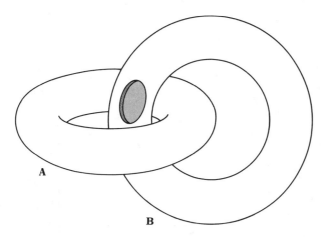

***Figure 57*** *Can torus B swallow torus A?*

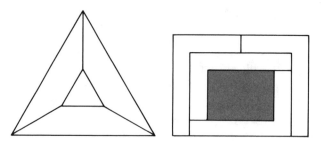

**Figure 58**   *A pair of tetrads*

The tetrad at the left in Figure 58 has no holes. Note that only three regions are congruent. The tetrad at the right has four congruent regions and a large hole. Is it possible, Buckley asked, to construct a tetrad with four congruent regions and no hole?

This question has been answered affirmatively by Scott Kim, a student at Stanford University. His results have not been published, and I am grateful to him for his permission to give some of them here.

Figure 59 shows a solution with four congruent hexagons. It is not known if a solution can be achieved with a polygon of fewer than six sides or if there is a solution with an outside border that is convex.

Part A of Figure 60 shows a solution with congruent polyhexes of order 4. (A polyhex is a union of congruent regular hexagons.) It is easy to show that no solution with lower-order polyhexes is possible.

Part B of the illustration shows a solution with congruent polyia-monds of order 10. (A polyiamond is a union of congruent equilateral

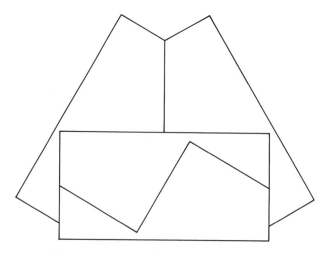

**Figure 59**   *A tetrad of congruent hexagons*

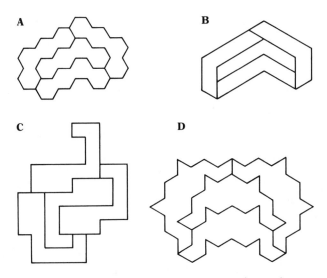

**Figure 60** *Solutions for a variety of tetrads*

triangles.) It is not known whether there is a solution with lower-order polyiamonds.

Part C of the illustration shows a solution with congruent polyominoes of order 12. (A polyomino is a union of congruent squares.) It is not known whether there is a solution with lower-order polyominoes.

Part D of the illustration shows a solution with congruent pieces that have bilateral symmetry and a bilaterally symmetric border. Is there such a solution with polygons of fewer sides?

It has been known since the 1870's that in three dimensions an infinite number of congruent solids can be put together so that every pair shares a common portion of surface. For readers who are unfamiliar with this result, here is the problem in terms of polycubes. (A polycube is a union of identical cubes.) Show how an infinite number of congruent polycubes can be put together, with no interior hole, so that every pair shares part of a surface. Such a structure proves that an unlimited number of colors are required for coloring any three-dimensional "map."

## 4. KNIGHTS AND KNAVES

Raymond Smullyan, a mathematician at the City University of New York whose remarkable chess problems are familiar to regular readers of my *Scientific American* column, is responsible for the following four charming logic puzzles involving knights and knaves and perhaps some other people. In all four problems a knight always tells the truth, a knave always lies.

> *A* says: "*B* is a knight."
>
> *B* says: "*A* is not a knight."
>
> Prove that one of them is telling the truth but is not a knight.
>
> *A* says: "*B* is a knight."
>
> *B* says: "*A* is a knave."
>
> Prove either that one of them is telling the truth but is not a
> knight or that one is lying but is not a knave.

In the above problems we must consider the possibility that a speaker is neither a knight nor a knave. In the next two problems each of the three people involved is either a knight or a knave.

> *C* says: "*B* is a knave."
>
> *B* says: "*A* and *C* are of the same type [both knights or both knaves]."
> What is *A*?
>
> *A* says: "*B* and *C* are of the same type."
>
> *C* is asked: "Are *A* and *B* of the same type?"
> What does *C* answer?

## 5. LOST-KING'S TOURS

Several years ago Scott Kim proposed what he calls the "Lost-King's Tour." This is a king's tour of a small chessboard subject to the following conditions:

First, the king must visit each cell once and only once.

Second, the king must change direction after each move; that is, it cannot move twice consecutively in the same direction.

Third, the number of spots where the king's path crosses itself must be minimized.

Part A of Figure 61 shows the only possible tour on a 3 × 3 board from cell *A* to cell *B*. It has one crossing and is unique, except of course for reflection by the main diagonal. A closed tour is impossible on this board. Closed tours with no crossings are easily found on the 4 × 4

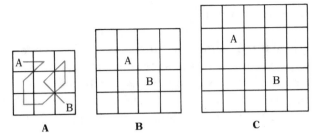

**Figure 61** *Lost-king's tours on three boards of different sizes*

board. On the 5 × 5 board a closed tour probably requires two crossings. The problem is less interesting on larger chessboards because crossing-free closed tours and crossing-free open tours from any cell to any other cell are believed to be always possible.

Here are two beautiful tour problems devised by Kim:

On the order-4 board shown in part B of the illustration find a lost-king's tour from *A* to *B* with as few as three crossings. The solution is unique.

On the order-5 board shown in part C of the illustration find a lost-king's tour from *A* to *B* with as few as two crossings. This problem is unusually difficult. Kim does not know whether the solution he found is unique or whether the problem can be solved with only one crossing.

## 6. STEINER ELLIPSES

This oldie goes back to Jakob Steiner, a noted Swiss geometer of the nineteenth century. My excuse for reviving it is that it is one of the best examples I know of a problem that is difficult to solve by calculus or analytic geometry but is ridiculously easy if approached with the right turn of mind and some knowledge of elementary plane and projective geometry.

We are given a 3, 4, 5 triangle. Its area is six square units. We wish to calculate both the smallest area of an ellipse that can be circumscribed around it and the largest area of an ellipse that can be inscribed inside it.

## 7. DIFFERENT DISTANCES

It is easy to place three counters on the cells of a 3 × 3 checkerboard so no two pairs of counters are the same distance apart. We assume that each counter marks the exact center of a cell and that distances are measured on a straight line joining the centers. Discounting rotations and reflections, there are five solutions, which appear in the top part of Figure 62.

It also is easy to put four counters on a 4 × 4 board so that all distances between pairs are different. There are 16 ways of doing it. On the 5 × 5 board the number jumps to 28.

In the January 1972 issue of the *Journal of Recreational Mathematics* Sidney Kravitz asked for solutions to the order-5 and the order-6 squares. The solution for the order-6 square proved to be difficult because for the first time the 3, 4, 5 right triangle (the smallest Pythagorean triangle) enters the picture. The fact that an integral distance of 5 is now possible both orthogonally and diagonally severely limits the patterns. As readers of the journal discovered, there are only two solutions. They appear in the bottom part of Figure 62.

For what squares of side *n* is it possible to place *n* counters so that all distances are different? As reported in the fall 1976 issue of the journal, John H. Muson proved (using a pigeonhole argument) that no solutions are possible on squares of order 16 and higher. Harry L. Nelson lowered the limit to 15. Milton W. Green, with an exhaustive-search computer program, established impossibility for orders 8 and 9 and found a unique solution for order 7. David Babcock, an editor of *Popular Computing*, confirmed these results through order 8. The problem was finally laid to rest in 1976 by Michael Beeler. His computer program confirmed the results indicated above and proved impossibility for orders 10 through 14.

The order-7 board, therefore, is the largest for which there is a solution. The solution is unique and extremely hard to find without a computer. Readers may nonetheless enjoy searching for it.

## 8. A LIMERICK PARADOX

An amusing variant of the old liar paradox appeared in the British monthly *Games and Puzzles*. It is presented here as the last of four "limericks."

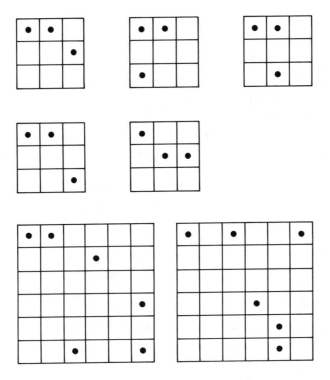

***Figure 62*** *A point-placement problem*

There was a young girl in Japan
Whose limericks never would scan.
   When someone asked why,
   She said with a sigh,
"It's because I always attempt to
get as many words into the last
line as I possibly can."

Another young poet in China
Had a feeling for rhythm much fina.
   His limericks tend
   To come to an end
Suddenly.

There was a young lady of Crewe
Whose limericks stopped at line two.

There was a young man of Verdun.

## ANSWERS

### 1. Pool-Ball Triangles

Except for reflection, the only solution of the problem with the 15 pool balls is shown in Figure 63. Col. George Sicherman of the State University of New York at Buffalo, who invented this problem, has found by computer that there are no solutions for similar triangles of orders 6, 7 and 8. Sicherman also found a simple parity proof of impossibility for all orders $2^n - 2$, where $n$ is greater than 2.

This is how the proof goes for order 6, the smallest case. Call the triangle's top row $a, b, c, d, e, f$. Since addition is the same as subtraction (modulo 2) we can express the other numbers (modulo 2) by adding. The second row is $a + b, b + c, c + d, d + e, e + f$. The next row begins $a + 2b + c, b + 2c + d. \ldots$ . Continue this way to the bottom number, which is $a + 5b + 10c + 10d + 5e + f$. The triangle contains 6 $a$'s, 20 $b$'s, 34 $c$'s, 34 $d$'s, 20 $e$'s and 6 $f$'s. All the numbers are even; therefore the triangle's parity is even. The triangle contains 11 odd and 10 even numbers, however, giving it an odd parity, so that we have a contradiction.

The above row of figures (6, 20, 34, 34, 20, 6) is the same as the seventh row of Pascal's triangle (1, 7, 21, 35, 35, 21, 7, 1) when the numbers are diminished by 1. Sicherman's general proof hinges on the well-known theorem that only rows numbered $2^n - 1$ of Pascal's triangle consist entirely of odd numbers.

Charles W. Trigg has proved, among other things, that every absolute-difference triangle of consecutive numbers must have 1 as its lowest number. He conjectures that there are no such triangles other than the 11 given last month and here. As a joke readers were asked to construct a difference triangle with 15 balls bearing the even numbers 2 through 30.

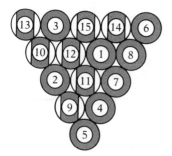

*Figure 63*
*Solution of the pool-ball problem*

The unique pattern is obtained at once by doubling each number in the solution shown.

## 2. Toroidal Cannibalism

One torus can be inside another in two topologically distinct ways: the inside torus may surround the hole of the outside torus or it may not.

If two toruses are linked and one has a "mouth," it cannot swallow the other so that the eaten torus is inside in the second sense. This result can be proved by drawing a closed curve on each torus in such a way that the two curves are linked in a simple manner. No amount of deformation can unlink the two curves. If one torus could swallow the other in the manner described, however, it could disgorge the eaten torus through its mouth and the two toruses would be unlinked. This result would also unlink the two closed curves. Because unlinking is impossible, cannibalism of this kind is also impossible.

The torus with the mouth can, however, swallow the other one so that the eaten torus is inside in the first sense explained above. Figure 64 shows how it is done. In the process it is necessary for the cannibal torus to turn inside out.

A good way to understand what happens is to imagine that torus *A* is shrunk until it becomes a stripe of paint that circles *B*. Turn *A* inside out through its mouth. The painted stripe goes inside, but in doing so it ends up circling B's hole. Expand the stripe back to a torus and you have the final picture of the sequence.

## 3. Exploring Tetrads

Figure 65 shows how two congruent polycubes (one shaded and one transparent) can be fitted together. By extension of the ends any finite number of such pieces can be nested in this manner so that every pair "touches" in the sense that they share a common surface and there are no interior holes. Extended to infinity, an infinite number of congruent polycubes, of order infinity, can mutually "touch."

If we drop the requirement of congruency but add the requirement of convexity, it has been known since about 1900 that an infinite number of noncongruent convex solids can mutually "touch." It is not known whether an infinite number of congruent convex solids can mutually touch, but Scott Kim has recently shown (although not published) how this arrangement can be achieved with an arbitrarily large finite number of such solids.

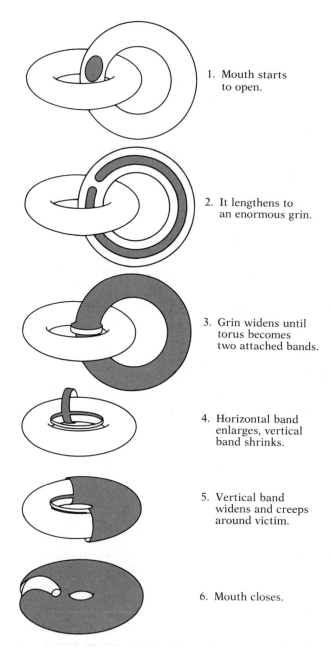

1. Mouth starts to open.

2. It lengthens to an enormous grin.

3. Grin widens until torus becomes two attached bands.

4. Horizontal band enlarges, vertical band shrinks.

5. Vertical band widens and creeps around victim.

6. Mouth closes.

*Figure 64* *How one torus eats another*

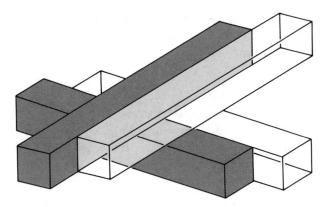

**Figure 65**  *Solution of the polycube problem*

## 4. Knights and Knaves

Raymond Smullyan's four logic problems are answered as follows:

1. *A* is either telling the truth or not. Suppose he is. Then *B* is a knight and telling the truth when he says *A* is not a knight. In this case *A* is telling the truth but is not a knight.

   Suppose *A* is lying. Then *B* is not a knight. *B*, however, is telling the truth when he says *A* is not a knight. Hence in this case *B* is telling the truth but is not a knight.

2. *B* is either telling the truth or not. Suppose he is. Then *A* is a knave and must be lying when he says *B* is a knight. In this case *B* is telling the truth but is not a knight.

   Suppose *B* is lying. Then *B* is surely not a knight; therefore *A* must be lying when he says *B* is a knight. Since *B* is lying, *A* is not a knave. In this case *A* is lying but is not a knave.

3. *B* is either a knight or a knave. Suppose he is a knight. *A* and *C* must then be the same type, as *B* says. *C* is lying when he says *B* is a knave; therefore *C* is a knave. If *C* is a knave, *A* must be also.

   Suppose *B* is a knave. Then *A* and *C* are different. *C* is telling the truth when he says *B* is a knave, so that *C* must be a knight. Because *A* and *C* are different, *A* must be a knave. In either case *A* is a knave.

4. Smullyan's solution of this problem is somewhat lengthy, and I shall content myself with a summary. *A* and *B* are either

knight-knight, knave-knave, knight-knave or knave-knight. In each case analysis shows that $C$, whether he is a knight or a knave, must answer yes.

### 5. Lost-King's Tours

Solutions to the two lost-king's tours are shown in Figure 66. The first tour is unique. The second is almost unique. One other pattern is obtained by modifying the path in the lower left corner as indicated by the dotted lines.

### 6. Steiner Ellipses

The ellipse problem was given for a 3, 4, 5 triangle, but we shall solve it for any triangle.

The largest ellipse that can be inscribed in an equilateral triangle is a circle, and the smallest ellipse that can be circumscribed around an equilateral triangle is also a circle. By parallel projection we can transform an equilateral triangle into a triangle of any shape. When that is done, the inscribed and circumscribed circles become noncircular ellipses.

Parallel projection does not alter the ratios of the areas of the triangle and the two closed curves; therefore the ellipses that result will have the maximum and minimum areas for any triangle produced by the projection. In other words, the ratio of the area of the smallest ellipse that can be circumscribed around any triangle to the area of the triangle is the same as the ratio of a circle to an inscribed equilateral triangle. Similarly, the ratio of the area of the largest ellipse that can be inscribed in any triangle to the area of the triangle is the same as the ratio of a circle to a circumscribed equilateral triangle.

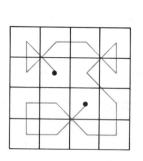 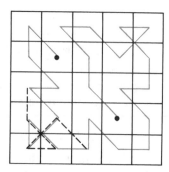

**Figure 66** *Solutions of lost-king's tours*

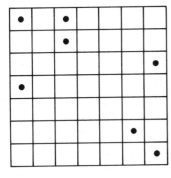

**Figure 67**
*Solution of the point-placement problem*

It is easy to show that the ratio of the inside circle to the triangle is $\pi/3\sqrt{3}$ and that the ratio of the outside circle to the triangle is four times that number. Applying this result to the 3, 4, 5 triangle, the area of the largest inside ellipse is $2\pi/\sqrt{3}$, and the area of the smallest outside ellipse is $8\pi/\sqrt{3}$.

Readers who would like to have a more formal proof will find it in Heinrich Dörrie's *100 Great Problems of Elementary Mathematics* (Dover, 1965), page 378 ff.

### 7. Different Distances

Figure 67 shows the only way (not counting rotations and reflections) to place seven counters on an order-7 matrix so that all distances between pairs of counters are different.

### 8. A Limerick Paradox

The paradox of the fourth limerick arises when the limerick is completed in one's mind: "Whose limericks stopped at line one." To complete it is to contradict what the limerick is asserting. The four paradoxical limericks prompted J. A. Lindon, the British comic versifier, to improvise the following new ones:

> A most inept poet of Wendham
> Wrote limericks (none would defend 'em).
>   "I get going," he said,
>   "Have ideas in my head.
> Then find I just simply can't."

> That things were not worse was a mercy!
>   You read bottom line first

Since he wrote all reversed—
He did every job arsy-versy.
A very odd poet was Percy!

Found it rather a job to impart 'em.
  When asked at the time,
  "Why is this? Don't they rhyme?"
Said the poet of Chartham, "Can't start 'em."

So quick a verse writer was Tuplett,
That his limerick turned out a couplet.

A three-lines-a-center was Purcett,
So when *he* penned a limerick (curse it!)
The blessed thing came out a tercet!

  Absentminded, the late poet Moore,
  Jaywalking, at work on line four,
    Was killed by a truck.

  So Clive scribbled only line five.

# ADDENDUM

The generalization of the pool-ball problem to triangles of order $n$, bearing consecutive numbers starting with 1, has been solved. Herbert Taylor found an ingenious way to prove that no TAD (triangle of absolute differences) could be made with triangular arrays of order 9 or higher. Computer programs eliminated TAD's of orders 6, 7 and 8; therefore the unique solution for the 15 pool balls is the largest TAD of this type.

Charles Trigg, in the paper "Absolute Difference Triangles," *Journal of Recreational Mathematics* (vol. 9, 1976–1977, pages 271–275), proves the uniqueness of the order-5 triangle and discusses some of its unusual characteristics, such as the five consecutive digits along its left side. On the basis of computer results and certain arguments he conjectures (correctly, as it turned out) that no absolute-difference triangles are higher than order 5. The only published proof known to me that the conjecture is true is given by G. J. Chang, M. C. Hu, K. W. Lih and T. C. Shieh in "Exact Difference Triangles," *Bulletin of the Institute of Mathematics*, Academia Sinica, Taipei, Taiwan (vol. 5, June 1977, pages 191–197).

Among other things, Trigg shows that any absolute-difference triangle of consecutive integers must begin with 1 and that each of the first $n$

consecutive integers (where $n$ is the triangle's order) must lie in a different horizontal row. It follows that the number at the triangle's bottom corner cannot exceed $n$.

Harry Nelson, who in 1977 edited the *Journal of Recreational Mathematics*, suggested a more general problem. Is there an absolute-difference triangle of order 5 that is formed with integers from the set 1 through 17? His computer program found 15 solutions. If the numbers are restricted to the interval 1 through 16, there are two solutions in addition to the 1-through-15 solution. With 8 missing, the top row is 5, 14, 16, 3, 15. With 9 missing, the top row is 8, 15, 3, 16, 14.

Solomon W. Golomb proposed three candidates for further investigation:

1. If all numbers in a TAD of order greater than 5 are distinct but not consecutive, how big is the largest number forced to be? (Example: An order-6 TAD is possible with the largest number as low as 22.)

2. Using all numbers from 1 to $k$, but allowing repeats, how big can $k$ be in a TAD of order $n$? (Example: An order-6 TAD is possible with $k$ as high as 20.)

3. For what orders is it possible to form a TAD modulo $m$, where $m$ is the number of elements in the triangle and the numbers are consecutive from 1 to $m$? Each difference is expressed modulo $m$. Such triangles can be rotated so that every element below the top row is the sum (modulo $m$) of the two numbers above it. Here, in rotated form, are the four order-4 solutions:

```
1 6 9 4       2 7 8 3       6 1 4 9       7 2 3 8
 7 5 3         9 5 1         7 5 3         9 5 1
  2 8           4 6           2 8           4 6
   0             0             0             0
```

A backtrack program by Golomb and Taylor found no solution for order 5. Col. George Sicherman, who invented the original pool-ball problem, reports a computer proof of impossibility for order 6. Higher orders remain open.

Robert Ammann, Greg Frederickson and Jean L. Loyer each found an 18-sided polygonal tetrad with bilateral symmetry (see Figure 68), thus improving on the 22-sided solution I had published.

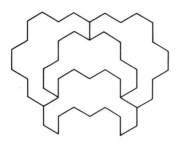

***Figure 68***
*A bilaterally symmetric tetrad with 18 sides*

Paul Erdös called my attention to a paper he wrote with Richard Guy, "Distinct Distances Between Lattice Points," *Elemente der Mathematik* (vol. 25, 1970, pages 121–133), in which they give the solution for the order-7 matrix and prove that no solutions exist for higher orders. They raise several related open questions, such as: What is the minimum number of lattice points that give distinct distances and are placed so that no point can be added without duplicating a distance?

When points are not confined to lattice points, we have a famous unsolved problem in combinatorial geometry. What is the smallest number of distinct distances determined by $n$ points on the Euclidean plane? Fan Chung, in her paper "The Number of Different Distances Determined by $n$ Points in the plane," *Journal of Combinatorial Theory* (series A, vol. 36, 1984, pages 342–354), cites earlier references going back to Erdös's proposal of the problem in 1946. A note added in proof gives the best-known lower bound as $n^{4/5}$. The bound will be discussed in a forthcoming paper by Chung, E. Szemerédi and W. T. Trotter.

When I ended the column with limericks of decreasing length, I referred to the one-line limerick as the "last of four." Draper L. Kauffman, John Little, John McKay, Thomas D. Nehrer and James C. Vibber were the first of many who told me I should have called it the last-but-one of five. The fifth, of course, has *no* lines, which is why other readers failed to notice it.

Tom Wright of Ganges, British Columbia, wrote: "I was interested in the limerick paradox, particularly in the decreasing two-line and one-line limericks. I wondered if you had, in fact, added the no-line limerick (about the man from Nepal), and I looked minutely to see if it wasn't there. On examination, my first impulse was to assume that it was indeed not there, since no space was provided, but further cogitation suggested that a no-line poem, requiring no space, might indeed be there. Unable to resolve this paradox by any logical proof, I am abjectly reduced to asking you whether or not a no-line limerick was not printed in the space not provided, or not."

# CHAPTER 10

Mathematical Induction
and Colored Hats

"Yes," he said, "a man has no need of eyes to
    perceive that."

—Plato, *The Republic*, Book V

Will the sun rise on January 1, 2001? There is no way to be
absolutely sure. It is always possible that the world will end before then
by divine decree or by some natural calamity. Perhaps a giant comet (as
in Immanuel Velikovsky's mythology) will cause the earth to stop rotating and the sun to stand still upon Gibeon. The most we can say is that it
is an excellent bet that on January 1, 2001, the sun will rise as usual.*
Our jump from a finite set of past sunrises to an infinite future set, or at
least to a future set with a large number of elements, is an empirical
induction.

---

*People like to say that some event is as certain as the sun rising tomorrow. "I like
that phrase," wrote Charles Peirce (*Collected Papers*, vol. 1, page 62), "for its great
moderation because it is infinitely far from certain that the sun will rise tomorrow."
There is not a single truth of science, Peirce said, on which he would "bet more than
about a million of millions to one."

Mathematicians have an analogous technique known as mathematical induction or complete induction that also supports a jump from a finite set of cases to a larger or an infinite number of cases. Unlike empirical induction, the mathematical technique is entirely deductive. A "jump proof," as it is sometimes called, is as certain as any proof can be in mathematics.

To prove something by mathematical induction we must first have a series of statements (usually an infinite series but not necessarily so) that can be put into a one-to-one correspondence with the sequence of positive integers. Second, we must establish that the statements are related to one another by what Bertrand Russell termed the "hereditary property." If any statement is true, its successor — the "next" statement — is true. Third, we must show that the first statement is true. It then follows with iron certainty that all the statements are true.

Jump proofs have been likened to a row of bricks or dominoes that are standing on end and all topple over when you unbalance the first one. Hugo Steinhaus compared mathematical induction to a pile of envelopes, each containing a note that says: "Open the next envelope, read the order and carry it out." If you are committed to obeying the order in the first envelope, you must open all the envelopes and obey all the orders.

Hundreds of classic problems in recreational mathematics are proved in the general case by mathematical induction. Into how many pieces can you cut a pie with $n$ straight cuts? In how few moves can you transfer $n$ disks in the Tower of Hanoi puzzle? In this chapter we discuss a class of mind-twisting logic puzzles for which the application of induction to the general case is less well known and is fraught with curious perils.

We begin with the old puzzle of the colored hats. Three men, $A$, $B$ and $C$, close their eyes while someone puts on the head of each either a black hat or a red one. They open their eyes. Each man sees the two hats not his own. If he sees a red hat, he raises his hand. As soon as he knows the color of his own hat, he must say so.

Suppose all three hats are red. The three men raise their hands. After a period of time $C$, who is smarter than the others, says: "My hat is red." How does he know?

$C$ reasons as follows. "Suppose my hat is black. $A$, seeing my black hat, will know at once that his own hat is red. Otherwise, why would $B$'s hand be raised? $B$ will reason the same way and also will know at once that his hat is red. Neither $A$ nor $B$, however, has said anything. Their hesitancy can only be explained if they see a red hat on me also. Therefore my hat is red."

Consider now the case of four men, all with red hats. If the fourth man, *D*, is smarter than the rest, he will reason: "Suppose my hat is black. The other three men have their hands raised because they see red hats. This is precisely the preceding case. After a suitable lapse of time, *C*, the smartest of the three, will deduce that his hat is red and say so." *D* then waits to see if *C* says anything. Because *C* says nothing, *D* knows his own hat is red.

Clearly this procedure generalizes. If there are five men. *E* will know his hat is red, because if it is black, the situation is reduced to the preceding case; after a suitable lapse of time *D* will know his hat is red. *D*'s silence proves to *E* that all hats, including his own, must be red. And so it goes for any number of men. Mathematical induction forces us to conclude that if *n* men all have red hats, the smartest of them will eventually deduce that his own hat is red.

This generalization usually provokes arguments because the problem demands so many fuzzy assumptions about degrees of smartness and lengthening lapses of time that the problem becomes unreal. Presumably if there are 100 men, after a few hours the smartest will know his hat is red, then after a while the second-smartest will know and so on down to the two stupidest men.

The fuzziness can be avoided by giving the same problem in a more precise form. There are three men, *A*, *B* and *C*, and five hats. Three hats are red and two are black. Each man is assumed to be honest and "rational" in the sense that he can quickly make any valid deduction no matter how complicated it is. As before, the men close their eyes, and an "umpire" puts a red hat on each man. The other two hats are hidden. Instead of being told to raise their hand if they see a red hat, the men are asked in order: "Do you know the color of your hat?"

*A* truthfully answers no. *B* also says no. *C* says, "Yes, my hat is red." How does he know?

A surprising aspect of this problem is that *C* can answer yes even though he is blind! Moreover, it is not necessary for *B* to see *A*'s hat. Think of the three men as being seated in a row of chairs, as is shown in Figure 69. Each man sees only the hats on the men in front of him. *C*, who is the man in the third chair, is blind in the sense that he sees no hats.

*C* reasons as follows: "*A* can say yes only if he sees two black hats. His saying no proves that hats *B* and *C* are not both black. Suppose my hat is

**Figure 69** *The problem of the colored hats* (The red hats are shown as gray)

black. *B* can see that it is. Therefore as soon as *B* hears *A* say no, he knows his own hat is red. (Otherwise hats *B* and *C* would be black and *A* would have said yes.) The fact that *B* also said no can be explained only if he sees my hat is red. Therefore I can answer yes."

This problem, like the preceding one, generalizes easily to *n* men seated in a row of chairs, a supply of *n* red hats and *n* − 1 black hats. Assume that a fourth man, *D*, is seated ahead of *C*. All hats are red. *D* reasons that if his hat is black, the three men behind him will see his black hat and know that only two black hats are left for themselves. Thus the problem is reduced to the preceding case, which is solved. After *A* and *B* had said no, *C* would say yes. But *C* also says no, which proves to *D* that his own hat must be red. Mathematical induction at once extends the solution to *n* men. If all have red hats, all will say no except the *n*th man, who will know that his hat is red.

A more difficult question can now be asked. Picture again three men in a row of chairs and assume that the umpire gives them any combination of hats from the set of five. The men are questioned in order of increasing "blindness" (*A*, *B*, *C*). Will one of them always be able to answer yes? And does this situation generalize to *n* men and a set of *n* red and *n* − 1 black hats? Will there always be a yes answer on or before the *n*th question regardless of what hats are put on their heads?

In most problems of this type we encounter a curious paradox. Consider the case of three men, all hats red and each man able to see the other two. *A* and *B* answer no, *C* answers yes. Why is it necessary to question *A*? Before *A* is asked both *B* and *C* know that he must say no; *B* knows because he sees the red hat on *C*, and *C* knows because he sees the red hat on *B*. If *B* and *C* know how *A* will answer, how can asking *A* and hearing his reply add any significant new information? On the other hand, if the questioning begins with *B*, *C* is unable to make his deduction. Can you explain this seeming paradox?

Hats of two colors are equivalent to hats labeled 0 and 1, the integers in binary notation. There are dozens of problems, closely related to the hat ones, in which more than two colors are involved but that are easier to understand if instead of colors we use positive decimal integers. The following two-person game was sent to me in 1976 by David Gale, a mathematician at the University of California at Berkeley.

The umpire chooses any pair of consecutive positive integers. A disk with one of the numbers is stuck to the forehead of one man and a disk with the other number is stuck to the forehead of the other man. Each man is honest and rational. Each one sees the other's number but not his own. Each knows (and knows the other knows) that the two numbers are consecutive.

The umpire asks each man if he knows his number, and the questioning continues back and forth until one man says yes. It is not hard to prove by the magic of induction that eventually the man with the higher number, $n$, will be the first to say yes, and that his yes will be in reply to question $n$ or $n - 1$. Readers are invited to analyze the game and to state under what conditions the high man says yes to question $n$ or to question $n - 1$. Only two variables need to be taken into account: whether the high man or the low man is asked first and whether the high number is odd or even.

The paradox of the hat game appears here in even more striking form. I paraphrase from Gale's letter. Assume that the numbers are 99 and 100 and that the man with 100 is asked first. He will say yes to the 100th question. But why ask the first two? Each man knows before the questioning begins that the first two answers must be no. How then can asking the first two questions furnish significant information? After the first two are asked, the men seemingly will know nothing they did not know before; therefore they should be no nearer to deducing their number than before, and the game will never end. How can a ritual intoning of no, which both men know must occur, shorten the number of questions required before a yes answer can be made? The argument appears impeccable.

Suppose we limit the integers to the counting numbers from 1 through 100. Each pair of consecutive numbers (1,2; 2,3; . . . ; 99,100) is written on a card. The umpire takes a card at random, puts its two numbers on the foreheads of two rational men and proposes the following game. The man with the lower number, $k$, must pay $k$ dollars to his opponent. The umpire asks $A$ if he wishes to play, and then he asks $B$. The payoff occurs only if both players say yes.

We now prove that a payoff never takes place. If $A$ sees 100, he knows he has 99, and so he says no. If he sees 99, he reasons: "I am 98 or 100. If I am 100, then $B$ (being rational) will say no and there will be no game. If I am 98, I surely should not play; therefore I must say no." If $A$ sees 98, he reasons: "I am 97 or 99. If I am 99, then $B$ will not play for the reasons given above. If I am 97, I shall lose; therefore I say no." And so on down even to seeing 1. If $A$ sees 1, he knows he will win, but he also knows that if he says yes, $B$ will say no.

Suppose the set of cards is infinite, with no upper bound on the integers. We now prove that both men will say yes. $A$ reasons: "I see the number $k$. My number is either $k - 1$ or $k + 1$. If I lose, I lose $k - 1$ dollars. If I win, I win $k$ dollars. It is equally probable that I shall win or lose, and since I stand to win more than I stand to lose, the game is in my favor. Naturally I agree to play." Of course $B$ reasons the same way. But this situation is preposterous because the game cannot favor both men.

The paradox can be magnified dramatically by taking a cue from J. E. Littlewood, who gives a version of the paradox in the first chapter of his *Mathematician's Miscellany*. Assume that there are $10^n$ duplicates of each card, where $n$ is the card's lower number. Thus 1,2 is on 10 cards, 2,3 is on 100 cards, 3,4 is on 1,000 cards, and so on. The game is played as before. If either player sees number $n$, he knows there are 10 times as many cards with $n + 1$ as there are with $n - 1$. Therefore in addition to a win being a dollar more than a loss, the probability of winning seems to be for each player 10 times greater than losing! Littlewood attributes this "monstrous hypothesis" to the physicist Erwin Schrödinger.

I shall not resolve this paradox because I am not sure just how to do it. It is not enough to prove the game fair with a playoff matrix. Obviously it is fair. The task is to explain what is wrong with the reasoning of $A$ and $B$.

John Horton Conway has given this game a confusing and deep generalization. It is easy to generalize to $n$ men and $n$ consecutive integers, but Conway does away with the consecutive proviso. We allow any non-negative integer (including 0) to be placed on the forehead of $n$ men, all honest and rational. On a blackboard, which all can see, are chalked $m$ different non-negative integers, just one of which is the sum of the numbers on the foreheads. Each man sees all the foreheads except his own. The umpire asks each man in turn, "Can you deduce the number on your head?" The questioning continues in cyclic order until a player concludes the game by saying yes.

Conway has proved the following remarkable theorem. If $m$, the number of sums on the blackboard, is not greater than $n$, the game must terminate. For example, suppose each man has a 2 and the blackboard sums are 6, 7 and 8. Conway asserts that the game ends with a yes to the 14th question.

Formulating a general algorithm for calculating when such games will terminate is, except for certain sets of numbers, extremely difficult and far from solved. Conway writes: "One gets into an infinite regress of the form '$A$ knows that $B$ knows that $C$ knows that $B$ knows that $C$ knows . . . ' even before the first question is asked, so it is very difficult to get a measure of the information available to each player. Indeed, I felt at one time that these considerations might make the game not well defined, and that we'd get into paradox trouble. Now I don't think so. I do know it's fatally easy to make mistakes in assessing what information is available."

Conway's game presents the same paradox we have considered before. In the given example it is easy to show that before the game starts each man can predict that the first three answers will be no, and so it seems as if asking these questions can be dispensed with because after the first round the men will be no better informed than before. If the first round were eliminated, however, the same argument would apply to the next round, and the game would never end.

Because mathematical induction often takes the form of "reducing to the preceding case," I close with an old joke. For a college freshman who cannot decide between physics and mathematics as his major subject the following two-part test has been devised. In the first part the student is taken to a room that contains a sink, a small stove with one unlighted burner and an empty kettle on the floor. The problem is to boil water. The student passes this part of the test if he fills the kettle at the sink, lights the burner and puts the kettle on the flame.

For part two the same student is taken to the same room, but now the kettle is filled and on the unlighted burner. Again the problem is to boil water. The potential physicist simply lights the burner. The potential mathematician first empties the kettle and puts it on the floor. This reduces the problem to the preceding case, which he has already solved.

## ANSWERS

Our problem concerned the game of three men, three red hats and two black hats. The men are seated in chairs so that *A* sees *B* and *C*, *B* sees only *C* and *C* sees no one. An umpire puts any three hats (from the set of five) on their heads. Each man is asked (in the order *A*, *B*, *C*) if he knows his hat's color. Will one of them always answer yes?

The answer is yes. An analysis of all color combinations will show that if the *ABC* order of hats is *RRR*, *RBR*, *BRR* or *BBR*, then *C* will say yes. If the order is *RRB* or *BRB*, then *B* and *C* will say yes. If it is *RBB*, all three will say yes. This analysis generalizes to *n* men with *n* red and *n* − 1 black hats. Consider *n* = 4. *D*, the "blind" man, reasons: "If my hat is black, the other three men will see it and know that only two black hats are left for them. The case will then be the same as the preceding one, which has been solved. If no one says yes, it can only be because my hat is red, and so I will say yes." And so on for any *n*. The first person to say yes is always the first one asked wearing a red hat and seeing no red hat.

John Erbland, a mathematics student at Northeastern University, thought of an amusing variant. Suppose there are *n* men, *n* − 1 black hats and only one red hat. Conditions are the same as before. Will the man with the red hat always say yes? If not, at what positions will he know the color of his hat?

The solution is curious. If *A* has the red hat, he will of course say yes because he sees *n* − 1 black hats. If *B* has the red hat, he will say no. Why? Because *A* must always answer yes; therefore his yes gives no information. If *C* has the red hat, he will say yes because he hears a yes from both *A* and *B*. If *D* has the red hat, he must say no because *C*'s yes could result from a deduction that he, *C*, has the red hat. This process generalizes by mathematical induction to the conclusion that the man with the red hat says yes if and only if he is in an odd position from the back. If he is in an even position, the yes of the man directly behind him fails to give sufficient information for a valid deduction.

The game of consecutive numbers on the foreheads of two men is analyzed as follows. Let *H* stand for the man with the higher number, *L* for the man with the lower one and *Qn* for the question number.

Suppose the numbers are 1 and 2. If *H* is asked first, he answers yes to *Q1*. If *L* is asked first, he says no. Then *H*, seeing the 1, says yes to *Q2*.

Suppose the numbers are 2 and 3. If *H* is asked first, he says no. *L* also says no. His answer proves that he does not see 1 on *H*, and so *H* knows that he is 3 and says yes to *Q*3. If *L* is asked first, he says no. As before, this answer tells *H* that his number is 3, and so he says yes to *Q*2.

Suppose the numbers are 3 and 4. If *H* is asked first, he says no. *L* also says no. *H* can now reason: "If I am 2, the game is reduced to the preceding case of 2, 3, with *L* asked first. Therefore *L* would say yes to *Q*2. Since *L* said no, it proves I am 4." Therefore *H* says yes to *Q*3. If *L* is asked first, he says no. *H* says no to *Q*2, which reduces the situation to the previous case, as before. *L*'s no to *Q*3 tells *H* that he is 4, and so he says yes to *Q*4.

Continuing in this way, the situation always reduces to a previously solved case. Let *n* be the higher number. *H* always wins. If he is asked first, he wins on *Q*(*n* − 1) if *n* is even and on *Qn* if *n* is odd. If he is asked second, he wins on *Qn* if *n* is even and on *Q*(*n* − 1) if *n* is odd. Note the curious fact that if the rule is that a player answer yes only if he knows he has the *lower* number, the game never ends!

Now for the seeming paradox. In the case of three men, all with red hats, it is true that both *B* and *C* know in advance that *A* will say no. But — and this is the easily overlooked and crucial point — *C* does not know before the first question is asked that *B* knows that *A* will answer no. Before the questioning starts, for all *C* knows his own hat may be black. If this is the case, *B* has no way of knowing whether *A* will say yes or no. Not until *A* says no does *C* know that *B* knew in advance of the question that *A* had to say no. Therefore the first question does add new information that is essential to *C*'s reasoning even though *C* knows in advance how *A* will answer.

Once this relation is understood it is not hard to see how the paradox is resolved for the Gale and Conway games. Suppose you are a player in Gale's game. Each new "No" in a game of arbitrary length gives you necessary new information in the general form of "I now know that you know that I know . . . that you don't know my number." It is the same for Conway's game. Each "No" provides each player with similar information about what the others know.

## ADDENDUM

I made no attempt to find the flaw in Littlewood's "monstrous" paradox about the infinite set of cards, each with a pair of consecutive integers, but more than 50 readers were quick to help me out. The paradox arises from the false assumption that numbers can be randomly selected from the infinite set of positive integers in such a way that all numbers are equally likely. There are many ways to demonstrate the absurdity of such a procedure. Choose any number $k$. The probability is 0 that a "randomly" selected positive integer is equal to or less than $k$, and the probability is 1 that it is greater than $k$. In other words, the chances are zero of selecting a number small enough to be written on any finite surface with symbols of finite size, no matter how small.

George Peter Wacktell made the point this way. You cannot shuffle an infinite deck of cards bearing all positive integers because if you could, the following contradictory theorem would hold: For any two cards taken from such a shuffled deck it is infinitely probable that each card bears a higher number than the other. In brief, the game in question cannot be played. Steven J. Brams, Elkan Halpern and Thomas Louis, Roger B. Lazarus, Frederick Mosteller and Herbert Robbins are among those who sent particularly detailed or interesting analyses of the problem.

Assume that the deck of cards is finite, so that an umpire can indeed randomize them and pick a pair of consecutive numbers small enough to be written. The upper bound of the pairs is not known. At least one player will always refuse to play. I. Martin Isaacs, a mathematician at the University of Wisconsin, amplified and corrected my earlier remarks about this game as follows:

"If the cards are numbered (1,2), (2,3), (3,4), . . . (possibly with some cards missing and others repeated, even allowing infinitely many cards and a method of random selection with a nonuniform probability distribution), then any player seeing an even number on his opponent should refuse to play. If he sees an odd number, he need not veto even (surprisingly) if the odd number he sees is known to be the highest number in the deck. This contradicts your analysis, which requires a person seeing the higher number to cast his veto even if the number is odd. He may agree to play, having confidence in the rationality of the other participant, who, seeing an even number, will not allow the payoff to occur.

"To prove my assertion, note that a player seeing 1 can only win and so need not veto. Thus it is rational play to veto on all numbers greater than 1 and play on seeing 1. (Of course, this is not the only rational way to play.) Each player, being rational, needs to protect himself against all possible rational strategies of his opponent. Thus if a player sees 2, he must veto to defend against the rational play of 'yes on 1, no on everything else' that his opponent might use. It now follows that 'yes on 3, no on everything else' is rational, since if a player sees 3, he knows he is safe if he has 2 because his opponent must veto when he sees 2. As a protection against the 'yes on 3 only' strategy, a rational player must veto on seeing 4, and this makes 'yes on 5 only' a rational plan. Continuing in this way, my assertion follows by induction.

"There is something very curious here: If the card (1,2) is missing from the deck, the above analysis remains unchanged. However, if the umpire announces that (1,2) is missing, then the game is equivalent to the preceding situation except now one must veto on odd numbers. On the other hand, if each player secretly looks through the deck and discovers that (1,2) is missing but does not know that his opponent also has this information, then rationality still demands vetoing evens (except for 2). It is again a question about whether '*A* knows that *B* knows. . . .' Throughout I am assuming that each player casts his veto by a secret ballot."

The literature of recreational mathematics abounds with paradoxical games that are related to Littlewood's paradox. Here is the simplest. Each player writes down a positive integer, and the person who writes the larger number wins a dollar. Each can reason: "Whatever number my opponent puts down, there will be only a finite number of smaller integers. Because there is an infinity of larger numbers, I am sure to win." Here again the fallacy lies in the impossibility of selecting a number at random from the infinite set of integers. There is an obvious limit to the size of a number one can write down in a finite time on finite sheets of paper. If the game were actually played, it would never end. Each player would keep scribbling digits on more and more sheets.

Suppose, however, we put an upper limit on the set of numbers. For example, each player has a spinner that selects a number from 1 through 100. Both players spin, and the person who gets the larger number must give that amount in dollars to the other player. Each can reason: "I am as likely to win as lose, but since I stand to win more than lose, the game is in my favor." Of course, each can also reason: "If I lose, I stand to lose more than I would win; therefore the game is *not* in my favor." The game obviously is fair, but it is not easy to say exactly what is wrong with either line of reasoning.

I first encountered this paradox in Maurice Kraitchik's *Mathematical Recreations* (Dover, 1953, pages 133–134), where it has the form of two strangers who agree to have the worth of their neckties evaluated. The person with the more expensive tie must give it to the other. In my *Aha! Gotcha* (W. H. Freeman and Company, 1982, page 106) I gave it the story line of two persons who compare the amounts of money in their wallets.

For a good analysis of this paradox, see Laurence McGilvery's 1987 article.

A more sophisticated version has recently been making the rounds, though I have not seen it in print. It goes like this. There are two players and two boxes. One box holds an unknown amount of dollars. The other contains twice that amount. A player is handed one box and told he may either keep it or exchange it for the other. He reasons: "The box I hold contains $x$ dollars. If I exchange it for the other box, I will get (with equal probability) $2x$ or $x/2$ dollars. The expected value after the exchange is half of $2x + x/2$, or $1.25x$. If I don't exchange, the expected value is $x$. Hence it is to my advantage to exchange." The conclusion is absurd, but what's wrong with the reasoning?

*Anno's Hat Tricks* (see the Bibliography), a delightful book for very bright children, is based entirely on mathematical induction puzzles about colored hats. The problems become increasingly difficult as you turn the pages, ending with the puzzle given here earlier about three players and five hats.

Many readers pointed out the similarity in the inductive reasoning about hats and the reasoning in a paradox known as the "unfaithful wives." The paradox seems to have first been given by George Gamow and Marvin Stern in their little book *Puzzle-Math* (Viking, 1958, pages 20–23). I cannot go into it here, but readers will find a version of it in my *Puzzles From Other Worlds* (Vintage, 1984, problem 37) and a full discussion in "Cheating Husbands and Other Stories" by Danny Dolev, Joseph Halpern and Yoram Moses, in an IBM Research Laboratory report of 1985, reprinted in the *Proceedings of the Fourth ACM Conference on Principles of Distributed Computing*, 1985.

## BIBLIOGRAPHY

*A Mathematician's Apology.* J. E. Littlewood. London: Methuen, 1953.

"Find the Needle." David Silverman, in *Journal of Recreational Mathematics*, 4, Problem 157, January 1971, p. 67; solution, 5, January 1972, p. 69; comment by Benjamin Schwartz, 5, October 1972, pp. 262–263.

"Using a Game as a Teaching Device." Charles Brumfiel, in *Mathematics Teacher*, 67, May 1974, pp. 386–391.

"A Calculus For Know/Don't Know Problems." A. K. Austin, in *Mathematics Magazine*, 48, January 1976, pp. 12–14.

"A Headache Causing Problem." John H. Conway and M. S. Patterson, in *Een Pak Met Een Korte Broek*. J. K. Lenstra et al., eds. Amsterdam, 1977.

*Anno's Hat Tricks*. Akihiro Nozaki. Illustrated by Mitsumasa Anno. Philomel, 1985.

"'Speaking of Paradoxes . . .' or Are We?" Laurence McGilvery, in *Journal of Recreational Mathematics*, 19, 1987, pp. 15–19.

"Computer Recreations: People Puzzles." A. K. Dewdney, in *Scientific American*, January 1989, pp. 106–109.

"The Conway Paradox: Its Solution in an Epistemic Framework." Peter van Emde Boas, Joroen Groenendijk, and Martin Stokhof, in *Formal Methods in the Study of Language*, J. A. G., Groenendijk, et al (eds.). My reprint gives no date for this book, identified only as MC Tract 135.

"More Paradoxes." David Gale, in *The Mathematical Intelligencer*, Vol. 16, No. 4, 1994, pp. 38–44.

# CHAPTER 11

## Negative Numbers

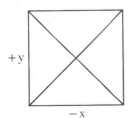

Inhabitants of Nega look surprisingly like us.
Their students seek a minus grade, they grumble
    at a plus.
In minus-fours the golfer walks, he never adds
    his score.
Meanwhile, non-minussed is his wife by prices at
    the store.

—IRVING E. FANG, *"A Tale of Star-crossed Lovers"*

When a child learns to talk, the names of the first few positive integers are almost as essential to his vocabulary as "dog," "cat" and "bird." Our primeval ancestors must have had a parallel experience. The counting numbers, sometimes called the natural numbers, were surely the first to be useful enough to require names. Today mathematicians apply the word "number" to hundreds of strange abstract beasts that are far removed from counting.

The first small step in enlarging the meaning of "number" was the acceptance of fractions as numbers. Although many things in the world are not commonly experienced as fractions (stars, cows, rivers and so on), it is easy to grasp the meaning of half an apple or a third of 12 sheep. But the next step, the acceptance of negative numbers, was so formidable that it was not until the seventeenth century that mathematicians began to feel truly comfortable about it. Many people are still uneasy with it, as is indicated by W. H. Auden's report of a jingle he was taught in school:

> Minus times minus equals plus.
> The reason for this we need not discuss.

One must distinguish negative numbers from subtraction. A child or an uneducated herdsman has no difficulty taking 6 cows from 10 cows. A "negative cow," however, is harder to imagine than a ghost cow. A ghost cow has at least some kind of reality, but a negative cow is less real than no cow. A cow from a cow leaves nothing, but adding a negative cow to a positive cow, causing both to vanish like a particle meeting its anti-particle, seems as ridiculous as the old joke about the individual whose personality was so negative that when he walked into a party, the guests would look around and ask, "Who left?"

That was how the ancient Greeks felt about negative numbers. They loved geometry and liked to think of mathematical entities as things they could diagram. "Numbers" were the counting numbers and the positive integral fractions that could be modeled with pebbles or spots on a slate. The Greeks' primitive algebra had no zero and no negative quantities. They were even reluctant to call 1 a number because, as Aristotle put it, numbers measure pluralities, and 1 is the measuring unit, not a plurality.

From B.C. by Johnny Hart. © 1988 Creators Syndicate, Inc.

It is important to realize that this attitude was to a large extent a matter of linguistic preference. Greek mathematicians knew that $(10 - 4)(8 - 2)$ equals $(10 \times 8) - (4 \times 8) - (2 \times 10) + (2 \times 4)$. To recognize such an equality is to accept implicitly what later was called the law of signs: The product of any two numbers with like signs is positive and the product of any two numbers with unlike signs is negative. It was just that the Greeks preferred not to call $-n$ a number. To them it was no more than a symbol for something to be taken away. You can take 2 apples from 10 apples, but taking 10 apples from 2 apples struck them as senseless. They knew that $4x + 20 = 4$ gives $x$ a value of $-4$, but they refused to write such an equation because its solution was "not a number." For the same reason $-\sqrt{n}$ was not recognized as a legitimate square root of $n$.

It is hard to know exactly how the earlier Babylonians regarded negative quantities, but they seem to have been more comfortable with them than the Greeks were. Chinese mathematicians before our era calculated rapidly with bamboo counting rods, using red rods for positive numbers (*cheng*) and black rods for negative numbers (*fu*). The same colors were later used for positive and negative written numerals. *Nine Chapters on the Mathematical Art*, a famous work of the Han period (roughly 200 B.c. to A.D. 200), explains the rod procedures and is believed to contain the first appearance in print of negative numbers as such. It does not, however, recognize negative roots or the law of signs.

A systematic algebra using zero and negative numbers did not develop until the seventh century, when Hindu mathematicians began to employ negative values for problems concerning credit and debts. Not only were they the first to use zero in a modern way but also they wrote equations in which negative numbers were symbolized by a dot or a tiny circle above the number. They explicitly formulated the law of signs and recognized that every positive number has two square roots, one positive and one negative.

Most Renaissance mathematicians in Europe, following the Greek tradition, viewed negative quantities with suspicion. Here again one must remember that it was more a matter of language preference than a failure to understand. Renaissance algebraists knew perfectly well how to manipulate negative roots; they just called them "fictitious roots." They knew perfectly well how to solve equations with negative numbers; they just avoided applying the word "number" to quantities less than zero.

By the seventeenth century a few bold mathematicians had altered their language to include negative numbers as legitimate numbers, but the practice continued to meet with resistance, sometimes from prominent mathematicians. Descartes spoke of negative roots as "false roots," and Pascal thought it nonsense to call anything less than zero a number. Pascal's friend Antoine Arnauld proved the absurdity of negative values as follows: The law of signs forces one to say that $-1/1 = 1/-1$. If this is taken as an equality between two ratios, we must assert that a smaller number is to a greater one as a greater number is to a smaller one. This seeming paradox, as Morris Kline points out in *Mathematical Thought from Ancient to Modern Times*, was much discussed by Renaissance mathematicians. Leibniz agreed that it was hard to resolve, but he defended negative numbers as useful symbols because they make possible correct calculations.

Some leading mathematicians of the seventeenth and eighteenth centuries—John Wallis and Leonhard Euler, to name two—accepted negative numbers but believed they were greater than infinity. Why? Because $a/0 = \infty$. Therefore if we divide $a$ by a number smaller than zero, say $-100$, must we not produce a negative quotient that exceeds infinity?

Symbols for addition and subtraction varied considerably during the Renaissance. Today's familiar plus and minus signs were first used in fifteenth-century Germany as warehouse marks. They indicated when a container held something that weighed over or under a standard weight. By the early sixteenth century German and Dutch algebraists were using $+$ and $-$ as operation signs, and the practice soon spread to England. Robert Recorde, a physician to Edward VI and Queen Mary, wrote a popular arithmetic in 1541 that was the first in English to use plus and minus signs, although not as operations. "Thys figure $+$, which betokeneth too much, as this lyne $-$, plaine without a cross lyne, betokeneth too lyttle" was how he explained them. A later book by Recorde was the first in England to use the modern sign for equality. "I will sette as I doe often in woorke use, a pair of paralleles . . . thus: $=$, because noe 2 thynges, can be moare equalle."

In the eighteenth century the algebraic use of negative numbers, identified by the minus mark, became common around the world. Nevertheless, most mathematicians remained discomfited. Their books included long justifications for the law of signs, and some authors went to extreme lengths rearranging equations to avoid multiplying two negative numbers. Here is a passage from *Dissertation on the Use of the Negative Sign in Algebra* (1758), by Baron Francis Masères, a British barrister who served as attorney general in Quebec:

> A single quantity can never be . . . considered as either affirmative or negative; for if any single quantity, as $b$, is marked either with the sign $+$ or with the sign $-$, without assigning some other quantity, as $a$, to which it is to be added, or from which it is to be subtracted, the mark

will have no meaning or signification: thus if it be said that the square of
−5 . . . is equal to +25, such an assertion must either signify no more
than that 5 times 5 is equal to 25 without any regard to the signs, or it
must be mere nonsense and unintelligible jargon.

The passage is quoted by Augustus De Morgan in *A Budget of Paradoxes*. Masères, De Morgan tells us, was such an honest lawyer that he
was not able to bear seeing his client victorious if he thought him guilty.
As a result, writes De Morgan, Masères's business gradually fell off.

A few pages earlier De Morgan makes a blistering attack on *The
Principles of Algebra*, by William Frend, a former clergyman who happened to be his father-in-law. (Frend's noisy banishment from Cambridge for his Unitarian views became a cause célèbre, passionately
championed by Samuel Taylor Coleridge and Joseph Priestley.) Frend's
two-volume work was probably the most ambitious algebra textbook
ever written in which zero and all negative numbers were as unwelcome
as Frend was at Cambridge.

De Morgan reprints in full Frend's hilarious burlesque of Rabelais, in
which Pantagruel gives a wild lecture on the uselessness of zero. A
plaintive footnote quotes Mrs. De Morgan: "[My father's] mental clear-
ness and directness may have caused his mathematical heresy, the rejec-
tion of the use of negative quantities in algebraical operations; and it is
probable that he thus deprived himself of an instrument of work, the use
of which might have led him to greater eminence in the higher
branches."

How can one do algebra without negative numbers? First one must
avoid any equation that leads to a negative number of real objects or
assigns negative magnitudes to them. Even when an equation leads to a
correct positive solution, it must be written so as to avoid a negative
value for an unknown. For example: When is a 29-year-old mother twice
as old as her 16-year-old daughter? We might write the problem as
$29 + x = 2(16 + x)$, then discover, perhaps to our surprise, that $x = −3$.
This result leads to the correct answer: The mother *was* twice as
old as her daughter when the mother was 26 and the daughter
was 13. An eighteenth century algebraist, if he was repelled by negative
numbers, would have avoided the −3 by rewriting the equation:
$29 − x = 2(16 − x)$. This arrangement gives $x$ the acceptable value of 3,
which of course leads to the same answer as before.

In past centuries, as it is today in beginning algebra classes, the main
stumbling block to the acceptance of negative numbers was "seeing"
how the product of two negative numbers can be positive. Positive times

positive offers no difficulty. Put three pairs of oranges in an empty bowl and the bowl will contain six oranges. Positive times negative begins to get mysterious but is not hard to understand if you grant the abstract reality of a negative orange. Put three pairs of negative oranges in the bowl and you have six negative oranges. But what on earth does it mean to multiply two negative oranges by −3? You have two ghostly oranges to start with, all less than nothing, and then you do something negative to them. Where do the six genuine oranges come from? They seem to appear in the bowl as a result of magic.

Trying to explain it by walking the number line, as is shown in Figure 70, does not get very far with beginning students. It is easy to identify positive integers with unit marks to the right of zero and negative integers with unit marks to the left. Addition is movement to the right and subtraction is movement to the left. To multiply 2 by 3 we go two units to the right and do this three times to arrive at 6. To multiply −2 by 3 we walk two units to the left and do this three times to arrive at −6. But what about −2 times −3? What paranormal force transports us abruptly from the left side of 0 to 6 on the right?

It is easy to forgive mathematicians of earlier centuries for regarding this concept as being preposterous. Indeed, the process was not fully understood until such abstract structures as groups, rings and fields were carefully defined. This is not the place to explain them, and so I content myself with pointing out that when mathematicians found it desirable to enlarge the concept of number to take in zero and negative numbers, they wanted the new numbers to behave as much like the old ones as possible.

One of the fundamental axioms of the old arithmetic is the distributive law, which states that $a(b + c) = ab + ac$; for example, $2(3 + 4) = (2 \times 3) + (2 \times 4)$. Change the 2 and 3 to negative numbers and the equality will still hold only if we adopt the rule that the product of two negative numbers is positive. If the product were negative, the equation would reduce to $-2 = -14$, a contradiction. In modern terms, the integers form a "ring" that is closed with respect to addition, subtraction and multiplication. This provision means that no matter how we add,

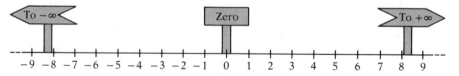

**Figure 70**   *The number line for integers*

subtract or multiply integers, whatever their sign, the result is always an integer. All the old laws of arithmetic for positive integers still hold, and we never encounter a contradiction. (We cannot always divide because we might get a fraction, and fractions are not elements of the ring.)

It is therefore not correct to say mathematicians can "prove" that the product of two negatives is a positive. It is rather a case of agreeing on rules that allow the negative numbers to obey all the old rules for counting numbers. If integral fractions are also included, the ring enlarges to a "field" that is closed under all four arithmetic operations.

Although there is no "proof" that $-2 \times -3 = 6$, it is easy to think of ways in which the law of signs applies to real situations. Indeed, it applies to all situations involving numbers on a scale that has two opposite directions or senses: east and west, up and down (as on a thermometer), forward and backward in time (or clockwise and counterclockwise on a clock), profit and loss and hundreds of others. It is because of these applications that the "signed numbers" are sometimes called "directed numbers."

In applying the law of signs to these examples we must always distinguish the quantities from the operations performed on them. The distinction is particularly necessary when we consider the multiplication of a signed quantity by a negative number. It is easy to understand what it means to take a positive or a negative quantity $n$ times, but what does it mean to take it $-n$ times? The clearest way to think about this mystifying operation is to break it into two parts:

1. Duplicate the quantity $n$ times.
2. Convert the result to its inverse with respect to zero. In other words, change the sign.

On the number line the second step is the same as reflecting a spot by a mirror on the zero mark. Imagine a bug at $-2$. To multiply its position by 3 there is no difficulty. We simply duplicate $-2$ three times to carry the bug to $-6$. But if the bug is at 2 and we wish to multiply by $-3$, our operation is to duplicate 2 three times, putting the bug at 6, and then invert. This procedure transports the bug to the mirror-image spot at $-6$. If the bug is at $-2$, multiplication by $-3$ operates the same way. We duplicate $-2$ three times, taking the bug to $-6$, and then inversion carries it to 6.

This may look like sorcery on the number line, but when we apply the procedure to many other situations, it seems quite normal. For example, suppose a man loses \$10 a day gambling. The future is defined as being positive and the past as being negative. Three days from now he will have lost \$30 ($3 \times -10 = -30$). Three days ago he had \$30 more than he has today ($-3 \times -10 = 30$). Equivalent situations arise on any directed scale. If water is sinking in a tank at a rate of three centimeters per minute, the level two minutes ago was $-3 \times -2 = 6$ centimeters higher. If the bug crawls west on the number line three centimeters per second, then two seconds ago it was $-3 \times -2 = 6$ centimeters east of its present position.

The most familiar property of objects that lends itself to negative magnitudes is weight. Add one-gram weights to your pockets and you are heavier. Attach helium-filled balloons to your body, each lifting with a force of one gram, and you are lighter. Remove three pairs of balloons and your weight increases by $-2 \times -3 = 6$ grams.

"Imagine a town where good people are moving in and out," wrote Roy Dubisch (*The Mathematics Teacher*, December 1971), "and bad people are also moving in and out. Obviously, a good person is + and a bad person −. Equally obvious, moving in is + and moving out is −. Still further, it is evident that a good person moving into town is a + for the town; a good person leaving town is a −; a bad person moving into town is a −; and, finally, a bad person leaving town is a +." If three pairs of bad people move out, the town gains $-2 \times -3 = 6$ points. One can model the situation with poker chips of two colors and a spot to represent the town.

Other models for teaching children the operations that can be done on the ring of integers have been proposed. Here is one of my own, so simple that others have surely thought of it before. It consists of a square board one centimeter thick in which 100 holes have been drilled in a square array. Each hole is fitted with a peg one centimeter long. The peg can be in one of three positions: flush with the board (0), projecting upward half a centimeter (+1) or projecting downward half a centimeter (−1). If all the pegs are flush, the board is in state 0. If $k$ pegs are up, it is in state $k$; if $k$ pegs are down, it is in state $-k$. (See Figure 71.)

To add $n$ to the state of the board, push $n$ pegs up, always pushing down pegs up (if there are any down pegs) before moving to flush pegs. To subtract $n$ from the board's state, push $n$ pegs down, taking first the up pegs (if there are any) before moving to flush pegs.

To multiply the state of the board by $n$, duplicate the state $n$ times. If the state is 0, there is nothing to do. If $k$ pegs are up, push up $n - 1$ more sets of $k$ pegs. If $k$ pegs are down, push down $n - 1$ more sets of $k$ pegs. To multiply the board's state by $-n$, first multiply by $n$ (as above) and then turn the board over.

A model that is probably easier to make is a board with little switches that can be in the up (+), middle (0) or down (−1) position. In this case

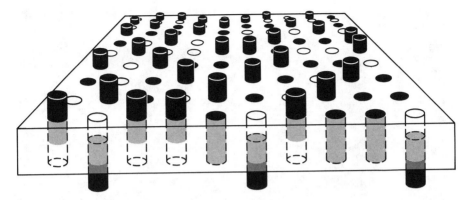

**Figure 71** *The pegboard that keeps track of signs*

multiplying by a negative number ends with rotating the board 180 degrees. (See Figure 72.)

If Aristotle were alive today and had 20 years to study modern algebra, he might still prefer to use "number" only for the counting numbers greater than 1. (There is a sense in which all "artificial numbers" are no more than constructions of the natural numbers.) This is one of those arguments over words that get nowhere. The point is that rings and fields, in which every element has its inverse or negative twin, are applicable to an enormous variety of natural objects and phenomena.

We may get into serious trouble trying to apply negative numbers and the law of signs to cubes and other "things" in the real world, as in J. A. Lindon's poem, (Figure 73), but sometimes the application is unexpectedly appropriate. Ignoring radiation, the pegboard is not a bad model of P. A. M. Dirac's famous theory about particles and antiparticles, a theory that predicted the existence of the positron!

## ADDENDUM

Lawrence Gilbert and Jack Bigelow, in a letter to *Mathematics Teacher* (vol. 79, February 1986), suggested a novel motion-picture model for multiplying by negatives. Imagine a movie of a car traveling down a road. Its movement forward along the road is positive; its backward motion, negative. Running the projector forward is a positive operation; running it backward is a negative operation. If the car is photographed moving forward along the road, multiplying it by positive corresponds to

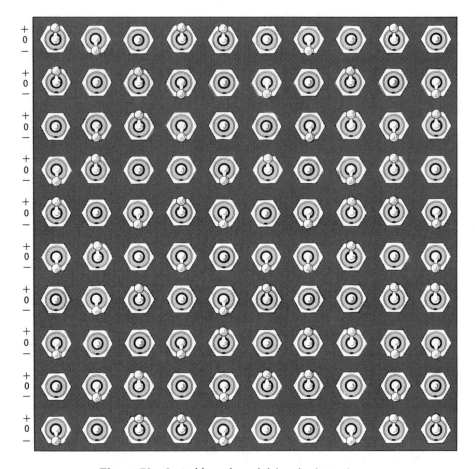

**Figure 72** *Switchboard model for the law of signs*

running the film forward to show the car still moving forward. Multiplying its motion by negative corresponds to running the same film backward, showing the car going backward. Now suppose the car was originally photographed moving backward down the road. Multiplying by positive is to run the film forward, which shows the car still going backward. To multiply by negative, the film is run backward, which shows the car moving forward.

Judith Kohn wrote to propose what strikes me as the best way to teach arithmetic operations to school children: a model that uses poker or bingo chips of two colors and an overhead projector. She later explained the model in a 1978 article (see the Bibliography).

A Positive Reminder

by J. A. Lindon

A carpenter named Charlie Bratticks,
Who had a taste for mathematics,
One summer Tuesday, just for fun,
Made a wooden cube side minus one.

Though this to you may well seem wrong,
He made it *minus* one foot long,
Which meant (I hope your brains aren't
   frothing)
Its length was one foot less than nothing.

In width the same (you're not asleep?)
And likewise minus one foot deep;
Giving, when multiplied (be solemn!),
Minus one cubic foot of volume.

With sweating brow this cube he sawed
Through areas of solid board;
For though each cut had minus length,
Minus *times* minus sapped his strength.

A second cube he made, but thus:
This time each one foot length was plus;
Meaning of course that here one put
For volume: *plus* one cubic foot.

So now he had, just for his sins,
Two cubes as like as deviant twins;
And feeling one should know the worst,
He placed the second in the first.

One plus, one minus—there's no doubt
The edges simply cancelled out;
So did the volume, nothing gained;
Only the surfaces remained.

Well may you open wide your eyes,
For these were now of double size,
On something which, thanks to his skill,
Took up no room and measured nil.

From solid ebony he'd cut
These bulky cubic objects, but
All that remained was now a thin
Black sharply-angled sort of skin

Of twelve square feet—which though not
   small,
Weighed nothing, filled no space at all.
It stands there yet on Charlie's floor;
He can't think what to use it for!

**Figure 73**   *On the mystery of 12 square feet*

Walter Penney defended the claim by Euler and Wallis that negative numbers exceed infinity. He suggested visualizing the number line as a circle of infinite radius, with infinity at the top, zero at the bottom, positive integers increasing in value counterclockwise from zero to infinity and negative numbers growing in the opposite direction from zero to infinity. If we define "greater than" to mean "farther in a counterclockwise direction," then the negative numbers lie on the far side of infinity.

Charles Rissanen sent a series of drawings to illustrate what happens in Lindon's poem. The carpenter starts with a large block of ebony. His negative cube is a hole cut out of this block. When his positive cube is placed inside the hole, the two cubes vanish and the original block is restored.

Confusion about multiplication frequently arises from a failure to distinguish quantities from operations. You can add two apples to three apples, but you can't multiply or divide two apples by three apples.

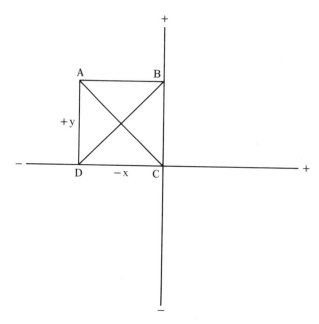

**Figure 74** *A square paradox*

Consider, for example, the following paradox that occurred to me. A square drawn on the Cartesian plane (Figure 74) has an area of $-(xy)$. By the Pythagorean theorem the diagonal *AC* is a positive number and similarly for diagonal *BD*. A familiar theorem of geometry says that the area of a square is half the product of its diagonals, and so if the diagonals are positive, the square's area must be positive, contradicting the prior result proving it negative. The fallacy springs from the fact that you can no more multiply by negative lengths than you can multiply by negative apples.

Richard Feynman, the eminent Cal Tech physicist, sent me a surprising document that has not, so far as I know, been published. I had pointed out in my column that although the real world does not contain such things as negative cows, negative numbers are useful in algebraic calculations about cows provided you don't end by asserting that so many negative cows are grazing in a pasture. Feynman's paper was a similar defense of the startling concept of negative probabilities. I cannot go into technical details, but essentially what he said was this: Situations arise in physics in which calculations are expedited by using negative probabilities provided you are careful not to assert that any actual event in the real world has a negative probability of occurring.

# BIBLIOGRAPHY

*A Budget of Paradoxes.* Augustus De Morgan. Open Court, Second Edition, 1915.

"Things and Un-things." W. W. Sawyer, in *Mathematics Teacher*, 51, January 1958, pp. 14–16.

"A Tale of Star-Crossed Lovers." Irving E. Fang, in *Uclan Review Magazine*, Spring 1960.

"A Physical Model for the Operations with Integers." Judith Kohn, in *Mathematics Teacher*, 71, December 1978, pp. 734–736.

"Impossible Numbers." Ernest Nagel, in *Teleology Revisited*, Chapter 8. Columbia University Press, 1979.

"On Multiplying Negative Numbers." Mary Crowley and Kenneth Dunn, in *Mathematics Teacher*, 78, April 1987, pp. 252–256.

"On Multiplying Negative Numbers." Mary Crowley and Kenneth Dunn, in *Mathematics Teacher*, 78, April 1985, pp. 252–256.

"Why Does a Negative Times a Negative Produce a Positive?" Vernon Thomas Sarver, Jr., in *Mathematics Teacher*, March 1986, pp. 178–80.

"Negative Membership." Wayne D. Blizard, in the *Notre Dame Journal of Formal Logic*, Vol. 31, Summer 1990, pp. 346–68.

CHAPTER **12**

# Cutting Shapes into N Congruent Parts

So the Wizard lost no more time, but leaping
forward he raised the sharp sword, whirled it
once or twice around his head, and then gave a
mighty stroke that cut the body of the sorcerer
exactly in two.
Dorothy screamed. . . .

—L. FRANK BAUM, *Dorothy and the Wizard in Oz*

A popular type of puzzle, often found in old puzzle books, is to divide a given shape into two, three or more equal parts. Sometimes "equal" means congruent; sometimes it means no more than equal in area. Readers of my earlier books may remember many problems of this type: enumerating the ways a chessboard can be cut along lattice lines into two or four congruent pieces, bisecting the yin-yang symbol (with one straight line) into four parts of equal area, dividing a square cake into *n* slices of equal volume, cutting "rep-tiles" into congruent copies of themselves and numerous others.

In this chapter we look at a variety of new problems of dividing shapes into equal parts. Some of them lead into significant areas of modern mathematics.

Let us begin with the simplest of the problems: cutting a plane shape into two, congruent parts. (Mirror images are considered congruent.) You might think that all such problems would be easy to solve, but they can be annoyingly difficult. As far as I know, there is no algorithm for deciding in general whether a shape can be divided into two or more congruent parts, and interesting theorems about such divisions are curiously scarce.

*Figure 75* *Shapes to be cut into congruent halves*

The reader is invited to try his skill at cutting each of the 12 shapes in Figure 75 into two identical pieces. There are no catches. Each half is connected simply (there are no holes or parts joined at single points). Some of these figures are from a French magazine column on recreational mathematics by Pierre Berloquin. (Scribner's has published translations of three of Berloquin's popular puzzle books.) The shape with the hole is difficult and suggests an unexplored region of puzzledom: shapes with holes that are to be cut into congruent pieces.

Note that the only kind of symmetry possessed by the fourth shape is bilateral, the axis of symmetry passing vertically down the middle. Any plane or solid figure with bilateral symmetry obviously can be sliced into congruent halves by cutting along its axis or plane of symmetry. Figure 76 shows how the Wizard of Oz bisected an evil Mangaboo into congruent parts while Eureka, Dorothy's pet cat, watched. (The scene occurs in the Glass City, below the earth's surface, where the inhabitants are vegetables.) Is it always necessary to cut along an axis or a plane of symmetry to divide a bilaterally symmetric figure into congruent parts? The answer is no, and the fourth shape proves it. You are asked to solve this figure in a way other than the obvious one.

Dividing a shape into $n$ congruent parts usually becomes more difficult as $n$ increases, particularly if the shape of the part is specified. The 12 pentominoes in Figure 77 pose an interesting quartering problem. (I am obliged to mention that "pentominoes" was registered as a trademark in 1975 by Solomon W. Golomb, who coined the term.) How many of the pentominoes can be dissected into four congruent pieces? All except three. I was surprised to discover that the nine with solutions can all be solved by using the same component shape—a smaller pentomino—and that this shape is unique. Can the reader discover it and identify the three impossible pentominoes?

Let us drop the requirement of congruence, demanding only that the $n$ parts have equal area. It seems intuitively clear that any plane shape can be cut into halves of equal area by a straight line. It is perhaps not so immediately obvious that the line can be parallel to any given line outside the figure. To prove this relation we make use of a famous theorem: If there is a continuous one-variable function in a closed interval from $A$ to $B$, then the function has a minimum value and a maximum value and all real values in between.

Let me give an example. You are walking up a mountain along a crooked path from $A$ to $B$. Your altitude at any moment is a continuous function of your position on the path. The theorem tells us that on the path there is at least one point (there may be more than one if the path

**Figure 76**  *The Wizard of Oz bisects a Mangaboo*

goes up and down) of minimum altitude, at least one point of maximum altitude and at least one point for every real value in between. The theorem seems obvious to the point of triviality, and yet it has fantastic power in proving theorems that are not at all obvious.

Consider the shaded region at the left in Figure 78. Outside it is an arbitrary line $x$. We want to prove that a line parallel to $x$ can be drawn through the figure that will exactly bisect its area. Imagine a line moving

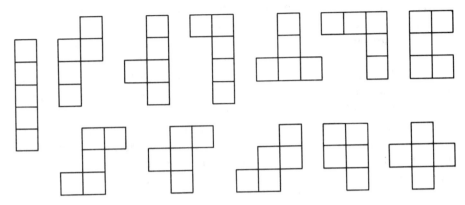

**Figure 77** *The 12 pentominoes*

slowly, always parallel to *x*, in the perpendicular direction indicated by the broken arrow. The line touches the region at *A* and leaves it at *B*. As the line moves over the region the area below it is a continuous function of the distance it has moved from *A*. The area of side *A* is zero at *A* and maximum at *B*. According to our theorem, somewhere in between is a point where the area is just half of the maximum. That is the point at which the line bisects the area.

The proof is so general that it applies not only to any connected shape, including one with holes, but also to disconnected regions. A glance at the right-hand part of Figure 78 should convince you that a line parallel to a given line can be drawn through any number of regions in

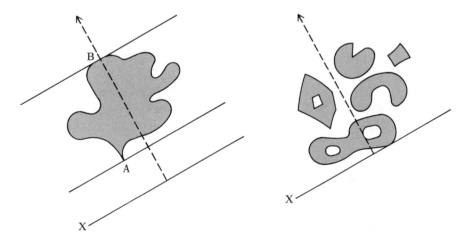

**Figure 78** *Proof of a theorem on bisecting an area*

such a way that the total area on one side of the line equals the total area on the other side.

Suppose you have two regions of any kind in the plane, as is shown in Figure 79. Is it always possible to draw a straight line that will simultaneously divide each region into parts of equal area? We can prove that it is. First bisect the white region with a line that misses the gray one. Rotate the line, always preserving the bisection of the white region. (We know we can do this from the above theorem.) As the illustration shows, the turning line will touch the gray region at point *A* and leave it at point *B*. As the line sweeps from *A* to *B* the area on side *A* will vary continuously from zero to the maximum. Hence there is a spot in between at which the area on side *A* is half of the total area.

We have proved for the plane a famous theorem that generalizes to all higher spaces. In 3-space the volumes of any three solids can be bisected by a plane; in 4-space the hypervolumes of any four figures can be bisected by a hyperplane of $N - 1$ dimensions. The 3-space theorem is sometimes called the "ham-sandwich theorem" because it applies to a generalized ham sandwich consisting of two pieces of bread and one piece of ham. No matter how the pieces are shaped or how they are placed in space, there is a plane that simultaneously halves all three.

The generalization to *N*-space requires high-powered mathematics, but in 2- and 3-space the proofs follow easily from our fundamental

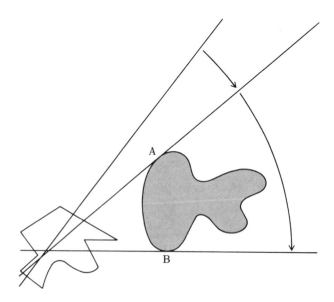

**Figure 79**   *Proving the ham-sandwich theorem in the plane*

theorem. Both proofs provide in a sense a way of finding a bisecting line or plane, but in practice it is not easy to find such a bisection. Connecting centroids (centers of gravity) will not do it because a straight line (or a plane) through the centroid of a plane (or a solid) figure does not necessarily bisect it into halves of equal area (or volume).

Given a simply connected figure, not necessarily convex, is there always a straight line that simultaneously bisects both the area and the boundary? There is, and the proof is much like the last one. Draw a line that bisects the boundary at $P$ and $Q$, as is shown in Figure 80. If this line, extended, also bisects the area, there is no more to do.

Assume that the line does not bisect the area. Attach to the line an arrow that points into the side of smaller area. Now move points $Q$ and $P$ clockwise around the boundary, always preserving the boundary's bisection. This maneuver will cause the line going through $P$ and $Q$ to alter its orientation in a continuous way. As it does so, the ratio of the areas on each side will vary continuously. After the line has rotated 180 degrees, it will coincide with its original position, but now the arrow will point into the larger of the two regions.

Consider the difference obtained by subtracting the area on the arrow side of the line from the other area. At the start it is positive. At the finish it is negative. This value clearly is a continuous function of the line's angle, and so there must be an angle at which the value is zero and the two areas are equal. The theorem also generalizes to higher spaces. A solid can be bisected in volume by a plane that bisects the surface area, and in general an $N$-space solid can be bisected in hypervolume by an $N - 1$ hyperplane that bisects its hypersurface.

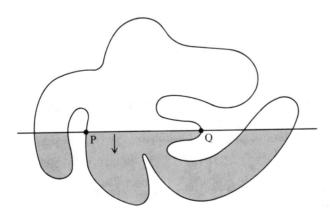

**Figure 80** *Proof of a theorem on bisecting a boundary and area*

Another beautiful theorem, equally far from obvious, states that any region—again it need not be connected simply or even connected—can always be exactly quartered by two perpendicular lines. The standard proof is both subtle and marvelous.

Imagine you have a sheet of transparent paper larger than the region. The sheet is divided into four quadrants by perpendicular lines $X$ and $Y$. The quadrant at the upper left is dark gray and the one at the upper right is light gray. Place the sheet over the region with the horizontal line below the region and the vertical line to the right. Slide the sheet up until line $X$ bisects the region, then slide the sheet to the left (letting $X$ slide along itself) until $Y$ likewise bisects the region. Label (on the region, not on the transparent sheet) the four parts of the figure $A$, $B$, $C$ and $D$. The arrangement is shown in the top part of Figure 81.

By construction $A$ plus $B$ equals $C$ plus $D$ and $A$ plus $C$ equals $B$ plus $D$. Subtracting the second equation from the first one results in $B - C = C - B$, or $2B = 2C$. Therefore $B$ equals $C$ and $A$ equals $D$.

If $A$ equals $B$, the region is quartered. Assume that it is not quartered and that region $B$ is larger than $A$. Therefore the area of the region's dark gray part, subtracted from the region's light gray part, is a positive number.

Rotate the sheet 90 degrees counterclockwise, always maintaining the double bisection of the region by $X$ and $Y$. Their intersection point may wander here and there. After the rotation is completed, the two lines necessarily return to their former position, as is shown in the bottom part of the illustration, except that now $X$ and $Y$ have exchanged places.

The light gray area of the region equals $A$. The dark gray area equals $C$, and because $C$ equals $B$, the dark gray area also equals $B$. Thus the areas of the dark gray and light gray regions have also exchanged places. Consequently if we subtract the larger dark gray area from the smaller light gray area, we now get a minus number.

It is easy to see how our fundamental theorem applies. The value obtained by taking the area of the dark gray region from the area of the light gray region is a continuous function of the angle of rotation as it varies from 0 to 90 degrees. Since the former value goes from plus to minus, there must be a value in between at which the difference is zero and the areas are equal. When that value is attained, the perpendicular lines precisely quarter the region.

This theorem also generalizes to higher spaces. Any solid can be divided into eight equal parts by three mutually perpendicular planes. In

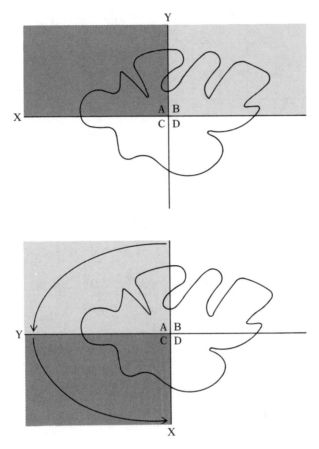

**Figure 81** *How any region can be quartered by two perpendicular lines*

general any N-dimensional solid can be cut into $2^n$ parts of equal "volume" by $n$ perpendicular "planes."

As in the preceding examples, the theorem is not much help in the practical problem of quartering nonsymmetric figures. We know, for instance, that a triangle of sides 3,4,5 can be quartered by two perpendicular lines, but the task of constructing those lines is an altogether different matter. Indeed, I know of no easy way to do it.

Figure 82 presents four equal-division problems for which the constructions are much less difficult than quartering the 3,4,5 triangle. They are nonetheless tricky enough to call for a considerable amount of ingenuity.

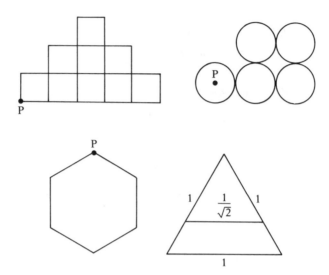

**Figure 82**  *Four problems in equal division*

The first problem is to divide the entire region of nine squares into two parts of equal area with a straight line that goes through corner *P*.

The second problem is to halve the total area of the five circles by drawing a straight line through *P*, which is the point at the center of the circle farthest to the left.

The third problem is to trisect the area of a regular hexagon with two straight lines that go through *P*.

The fourth problem is to bisect an equilateral triangle with a curve of minimum length. The illustration of the triangle shows the shortest bisecting straight line, but there is a bisecting curve that is shorter.

## ANSWERS

Figure 83 shows how the 12 shapes given in Figure 75 can be divided into congruent halves. Figure 84 shows how nine of the 12 pentominoes can be dissected into the same four congruent parts. The three blank pentominoes cannot be cut into four congruent parts of any shape.

Figure 85 answers the four problems at the end of the chapter. To bisect the nine squares draw the 10th square shown with broken lines. Rule *AB* to get point *C*, then join *P* to *C*. If the squares have sides of length 1, then *CD* equals 1/4 and it is easy to see that *PC* bisects the

**Figure 83** *Answers to bisection problems*

original figure. To bisect the five circles add three additional circles as shown by the broken lines. The line through the centers of two circles obviously halves the total area. (Both problems are from *A Problem a Day*, by R. M. Lucey, Penguin Books, 1937.)

The hexagon at the bottom is trisected by joining P to C and D, the midpoints of two sides. Assume that the equilateral triangles have areas

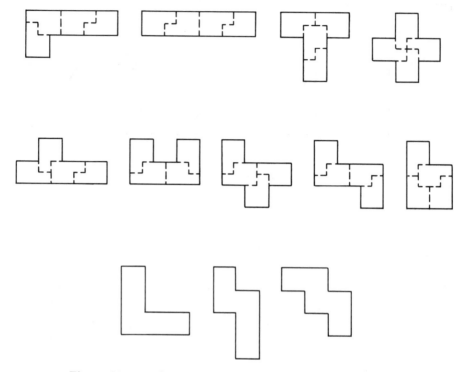

**Figure 84**   *Dividing pentominoes into four congruent parts*

of 1. The area of *PAB* is 1; therefore the area of *PBE* is 2 and the rest follows. I was unable to find any comparably simple way to trisect a regular pentagon with a line through a corner.

The middle two hexagons show how Leo Moser proved that the minimum-length curve bisecting an equilateral triangle is the arc of a circle. Whatever the shape of the bisecting curve, it will form a closed curve if the triangle is reflected around one vertex as is shown. Such a curve cuts the hexagon in half, and it has a fixed area. The figure of minimum perimeter that encloses a given area is the circle; therefore the minimum-length bisecting curves inside each triangle are arcs of a circle. (This exercise is from *Mathematical Quickies*, by Charles W. Trigg, McGraw-Hill, 1967.)

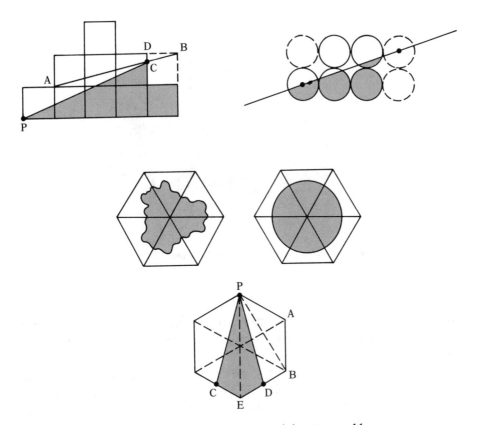

**Figure 85**  *Solutions to four equal-division problems*

## ADDENDUM

My solution to the bisection of the five circles required the construction of a sixth circle. Many readers sent in a construction that doesn't use a sixth circle (Figure 86). It is obvious from the illustration that the solid line bisects the five circles. Max Gordon Phillips, of Sunnyvale, California, then improved the improvement by eliminating *all* prior constructions. As he pointed out, a line drawn through the point where two circles touch will bisect the area of the two circles regardless of how it is rotated. If this line also passes through the center of a third circle, with the remaining two circles on opposite sides of the line, all five circles

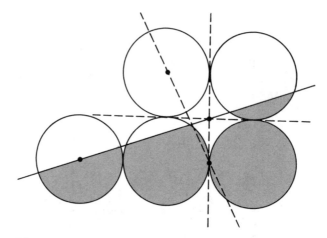

**Figure 86** *Improved solutions to a bisection problem*

will have their total area bisected. There are five such lines, each passing through the center of one of the circles. One is shown by the dotted line in the illustration.

I stated that I knew no easy way to trisect a pentagon with lines extending from one corner. Carl F. Von Mayenfeldt of Vancouver supplied the trisection shown in Figure 87. Assume that the pentagon has sides of length 1.

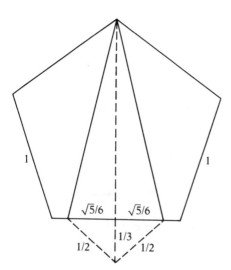

**Figure 87**
*Trisecting the area of a pentagon*

After proving that a pair of perpendicular lines divide the area of any figure into quarters, I called attention to the difficulty of constructing such a pair of lines for the right triangle of sides 3,4,5. Von Mayenfeldt was the first to send the solution and to prove it unique. The same construction, too complicated to give here, was later sent by Windschitl and Cecil G. Phipps.

Although no reader found a general algorithm for solving all problems that involve cutting any shape into two congruent halves (when possible), many readers found procedures that apply to many such problems. For example, such problems can sometimes be solved by tracing the shape to be divided on transparent paper and then turning the tracing to intersect the original figure in various ways. Abel Bomberault, Robert W. Davis, James R. Fienup, David Fleming, Allen J. Schwenk, Daniel Sleator, Laszlo M. Vesei and Marcel Vinokur were the first to send suggestions of this kind.

Don Taylor proposed an amusing bet derived from the fixed-point theorem on which the possibility proofs I gave were based. Bet someone you can draw a line *exactly* 1 inch long. You win by drawing a line about 3 inches long. Because the length varies continually from less than an inch to more than an inch, there must be a point along the line where it is precisely 1 inch from that point to an end.

We saw how two figures on the plane can always be bisected by a straight line. In *Mathematical Snapshots* (Third Edition, Oxford University Press, 1969, page 145) Hugo Steinhaus asserts that any three figures on the plane can always have their combined areas bisected by a suitably placed circle. No proof is given, but a footnote refers to a paper by Steinhaus in *Fundamenta Mathematica* (vol. 33, 1945, pages 245–263).

Since this chapter first appeared in *Scientific American*, many problems of cutting a figure into *n* congruent parts have appeared in mathematical journals and popular magazines such as *Games* (published here by *Playboy*) and *Juegos* and *Cacumen*, two Argentine periodicals. Springer-Verlag's *Mathematics Calendar* for 1979 featured an essay on such tasks and a page of nine extremely difficult bisection puzzles.

In Figure 88 I have a sampling of bisection problems. The grid lines are there only to make clear the shape, not to suggest that the bisecting line must be confined to the grid. The last figure, invented by S. W. Golomb, is the hardest. As a hint, I will say that in this case (as in some others) the bisecting line does follow the grid.

The top illustration of Figure 89 is one of three trisection problems in the British journal *Eureka* (Spring 1984). Show how to cut it into three congruent polyominoes. The bottom illustrations are quartering prob-

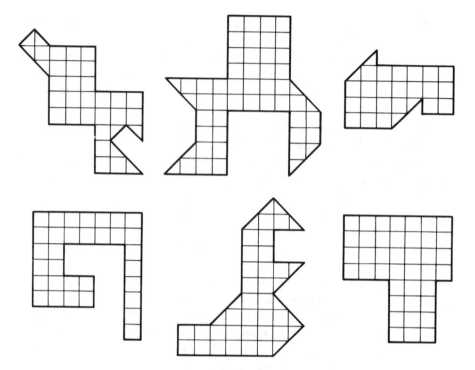

**Figure 88**   *More bisection problems*

lems from Manjunath Hegde of India. The first shows how a figure is easily divided into four congruent triangles. Your task is to divide it a different way into four congruent parts, each containing one of the four spots. Hegde's second puzzle is to divide the figure into four congruent shapes. In all three problems the divisions are made along grid lines. It would diminish the solving fun if I provided answers to these three tasks, as well as answers to the six bisection problems.

I find in my files a newspaper clipping of January 21, 1983. An Associated Press dispatch from Central City, Kentucky, reports a judge's restraining order barring Virgil M. Everhart from bisecting his house. It seems that his wife's divorce action demanded an equal property settlement. Mr. Everhart, using drills and saws, began slicing his house into two equal parts which he called "His" and "Hers."

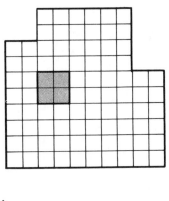

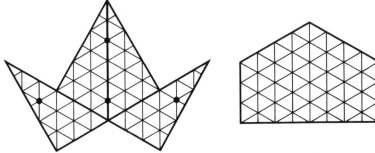

**Figure 89**   *Trisection problem* (top) *and quartering problems* (bottom)

## BIBLIOGRAPHY

"Functions and Limits." Richard Courant and Herbert Robbins, in *What is Mathematics?* Oxford University Press, 1941.

"Generalized 'Sandwich' Theorems." A. H. Stone and J. W. Tukey, in *Duke Mathematical Journal*, 9, June 1942, pp. 356–359.

"Cuts and Congruence." *Mathematics Calendar*. Springer-Verlag, 1979.

"On a Quadrisection Problem of M. Gardner." Sin Hitotumatu, in *Research Activities*, Vol. 4, 1992, pp. 1–4. A construction is given for the quadrisection of any right triangle by two perpendicular lines.

"Geometric Dissections." Nob Yoshigahara, in *Puzzle World*, Vol. 1, Summer 1992, pp. 48–51.

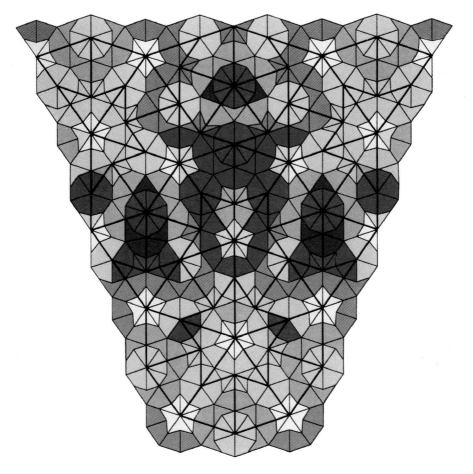

**Plate 1**  *The king's empire*

***Plate 2*** *A computer generated earthlike planet, seen above an imaginary moonscape*

**Plate 3**  *Members of the Oulipo*
*Standing, left to right: Jean Fournel, Michèle Métail, Luc Etienne, Georges Perec, Marcel Renabou, Jean Lescure, Jaques Duchateau*
*Seated, left to right: Italo Calvino, Harry Mathews, François Le Lionnais, Raymond Queneau, Jean Queval, Claude Berge*

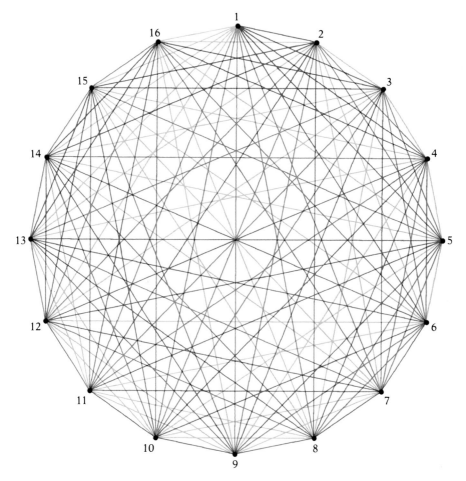

**Plate 4**   *A critical three-coloring of $K_{16}$, a complete graph on 16 points*

# CHAPTER 13

## Trapdoor Ciphers

> Few persons can be made to believe that it is not
> quite an easy thing to invent a method of secret
> writing which shall baffle investigation. Yet it
> may be roundly asserted that human ingenuity
> cannot concoct a cipher which human ingenuity
> cannot resolve.
>
> —EDGAR ALLAN POE

The upward creep of postal rates accompanied by the deterioration of postal service is a trend that may or may not continue, but as far as most private communication is concerned, in a few decades it probably will not matter. The reason is simple. The transfer of information will surely be much faster and much cheaper by "electronic mail" than by conventional postal systems. Before long it should be possible to go to any telephone, insert a message into an attachment and dial a number. The telephone at the other end will print out the message at once.

Government agencies and large businesses will presumably be the first to make extensive use of electronic mail, followed by small businesses and private individuals. When this starts to happen, it will become increasingly desirable to have fast, efficient ciphers to safeguard information from electronic eavesdroppers. A similar problem is involved in protecting private information stored in computer memory banks from snoopers who have access to the memory through data-processing networks.

It is hardly surprising that in recent years a number of mathematicians have asked themselves: Is it possible to devise a cipher that can be rapidly encoded and decoded by computer, can be used repeatedly without changing the key and is unbreakable by sophisticated cryptanalysis? The surprising answer is yes. The breakthrough is scarcely two years old, yet it bids fair to revolutionize the entire field of secret communication. Indeed, it is so revolutionary that all previous ciphers,

---

A CIPHER THAT DEFEATED POE

GE JEASGDXV,

ZIJ GL MW, LAAM, XZY ZMLWHFZEK EJLVDXW KWKE TX LBR ATGH LBMX AANU BAI VSMUKKSS PWN VLWK AGH GNUMK WDLNZWEG JNBXVV OAEG ENWB ZWMGY MO MLW WNBX MW AL PNFDCFPKH WZKEX HSSF XKIYAHUL. MK NUM YEXDM WBXY SBC HV WYX PHWKGNAMCUK?

In 1839, in a regular column Edgar Allan Poe contributed to a Philadelphia periodical, *Alexander's Weekly Messenger*, Poe challenged readers to send him cryptograms (monoalphabetic substitution ciphers), asserting that he would solve them all "forthwith." One G. W. Kulp submitted a ciphertext in longhand. It was printed as shown above in the issue of February 26, 1840. Poe "proved" in a subsequent column that the cipher was a hoax — "a jargon of random characters having no meaning whatsoever."

In 1975 Brian J. Winkel, a mathematician at Albion College, and Mark Lyster, a chemistry major in Winkel's cryptology class, cracked Kulp's cipher. It is not a simple substitution — Poe was right — but neither is it nonsense. Poe can hardly be blamed for his opinion. In addition to a major error by Kulp there are 15 minor errors, probably printer's mistakes in reading the longhand.

Winkel is an editor of a new quarterly, *Cryptologia*, available from Albion College, Albion, MI 49224. The magazine stresses the mathematical and computational aspects of cryptology. The first issue (January 1977) tells the story of Kulp's cipher and gives it as a challenge to readers. So far only three readers have broken it. I shall give the solution in the answer section.

together with the techniques for cracking them, may soon fade into oblivion.

An unbreakable code can be unbreakable in theory or unbreakable only in practice. Edgar Allan Poe, who fancied himself a skilled cryptanalyst, was convinced that no cipher could be invented that could not also be "unriddled." Poe was certainly wrong. Ciphers that are unbreakable even in theory have been in use for half a century. They are "one-time pads," ciphers that are used only once, for a single message. Here is a simple example based on a shift cipher, sometimes called a Caesar cipher because Julius Caesar used it.

First write the alphabet, followed by the digits 0 through 9. (For coding purposes 0 represents a space between words, and the other digits are assigned to punctuation marks.) Below this write the same sequence cyclically shifted to the right by an arbitrary number of units, as is shown in Figure 90. Our cipher consists in taking each symbol in the plaintext (the message), finding it in the top row and replacing it with the symbol directly below it. The result is a simple substitution cipher, easily broken by any amateur.

In spite of its simplicity, a shift cipher can be the basis of a truly unbreakable code. The trick is simply to use a different shift cipher for each symbol in the plaintext, each time choosing the amount of shift at random. This is easily done with the spinner shown in Figure 91. Suppose the first word of plaintext is THE. We spin the arrow and it stops on K. This tells us to use for encoding T a Caesar cipher in which the lower alphabet is shifted 10 steps to the right, bringing A below K as is shown in the illustration. T, therefore, is encoded as J. The same procedure is followed for every symbol in the plaintext. Before each symbol is encoded, the arrow is spun and the lower sequence is shifted accordingly. The result is a ciphertext starting with J and a cipher "key" starting with K. Note that the cipher key will be the same length as the plaintext.

To use this one-time cipher for sending a message to someone — call him Z — we must first send Z the key. This can be done by a trusted courier. Later we send to Z, perhaps by radio, the ciphertext. Z decodes

```
A B C D E F G H I J K L M N O P Q R
0 1 2 3 4 5 6 7 8 9 A B C D E F G H

S T U V W X Y Z 0 1 2 3 4 5 6 7 8 9
I J K L M N O P Q R S T U V W X Y Z
```

**Figure 90**
*A Caesar cipher with a 10-shift*

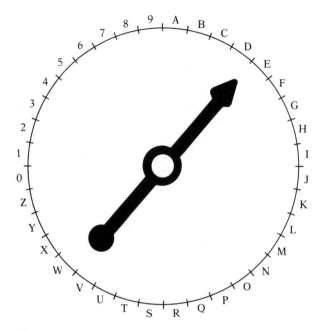

**Figure 91** *Randomizer for encoding a "one-time pad"*

it with the key and then destroys the key. The key must not be used again because if two such ciphertexts were intercepted, a cryptanalyst might have sufficient structure for breaking them.

It is easy to see why the one-time cipher is uncrackable even in principle. Since each symbol can be represented by any other symbol, and each choice of representation is completely random, there is no internal pattern. To put it another way, any message whatever having the same length as the ciphertext is as legitimate a decoding as any other. Even if the plaintext of such a coded message is found, it is of no future help to the cryptanalyst because the next time the system is used the randomly chosen key will be entirely different.

One-time pads are in constant use today for special messages between high military commanders, and between governments and their high-ranking agents. The "pad" is no more than a long list of random numbers, perhaps printed on many pages. The sender and receiver must of course have duplicate copies. The sender uses page 1 for a cipher, then destroys the page. The receiver uses his page 1 for decoding, then destroys his page. When the Russian agent Rudolf Abel was captured in New York in 1957, he had a one-time pad in the form of a booklet about the size of a postage stamp. David Kahn, who tells the story in his marvelous history *The Codebreakers*, says that the one-time pad is the standard method of secret radio communication used by the U.S.S.R. The famous "hot line" between Washington and Moscow also makes use of a one-time pad, the keys being periodically delivered through the two embassies.

If the one-time pad provides absolute secrecy, why is it not used for all secret communication? The answer is that it is too impractical. Each time it is employed, a key must be sent in advance and the key must be at least as long as the anticipated message. "The problem of producing, registering, distributing and canceling the keys," writes Kahn, "may seem slight to an individual who has not had experience with military communications, but in wartime the volumes of traffic stagger even the signal staffs. Hundreds of thousands of words may be enciphered in a day; simply to generate the millions of key characters required would be enormously expensive and time-consuming. Since each message must have its unique key, application of the ideal system would require shipping out on tape at the very least the equivalent of the total communications volume of a war."

Let us qualify Poe's dictum by applying it only to ciphers that are used repeatedly without any change in the key. Until recently all cipher systems of this kind were known to be theoretically breakable provided the code breaker has enough time and enough ciphertext. Then in 1975 a new kind of cipher was proposed that radically altered the situation by supplying a new definition of "unbreakable," a definition that comes from the branch of computer science known as complexity theory. These new ciphers are not absolutely unbreakable in the sense of the one-time pad, but in practice they are unbreakable in a much stronger sense than any cipher previously designed for widespread use. In principle these new ciphers can be broken, but only by computer programs that run for millions of years!

The three men responsible for this remarkable breakthrough are Whitfield Diffie and Martin E. Hellman, both electrical engineers at Stanford University, and Ralph Merkle, then an undergraduate at the University of California, Berkeley. Their work was partly supported by the National Science Foundation in 1975 and was reported by Diffie and Hellman in their 1976 paper "New Directions in Cryptography". In it Diffie and Hellman show how to create unbreakable ciphers that do not require advance sending of a key or even concealment of the method of encoding. The ciphers can be efficiently encoded and decoded, they can be used over and over again and there is a bonus: The system also provides an "electronic signature" that, unlike a written signature, cannot be forged. If Z receives a "signed" message from A, the signature proves to Z that A actually sent the message. Moreover, A's signature cannot be forged by an eavesdropper or even by Z himself!

These seemingly impossible feats are made possible by what Diffie and Hellman call a trapdoor one-way function. Such a function has the following properties: (1) it will change any positive integer x to a unique positive integer y; (2) it has an inverse function that changes y back to x; (3) efficient algorithms exist for computing both the forward function and its inverse; (4) if only the function and its forward algorithm are known, it is computationally infeasible to discover the inverse algorithm.

The last property is the curious one that gives the function its name. It is like a trapdoor: easy to drop through but hard to get up through. Indeed, it is impossible to get up through the door unless one knows where the secret button is hidden. The button symbolizes the "trapdoor information." Without it one cannot open the door from below, but the button is so carefully concealed that the probability of finding it is practically zero.

Before giving a specific example, let us see how such functions make the new cryptographic systems possible. Suppose there is a group of businessmen who want to communicate secrets to one another. Each devises his own trapdoor function with its forward and backward algorithms. A handbook is published in which each company's encoding (forward) algorithm is given in full. The decoding (inverse) algorithms are kept secret. The handbook is public. Anyone can consult it and use it for sending a secret message to any listed company.

Suppose you are not a member of the group but you want to send a secret message to member Z. First you change your plaintext to a long number, using a standard procedure given in the handbook. Next you look up Z's forward algorithm and your computer uses it for rapid encoding of the ciphertext. This new number is sent to Z. It does not matter at all if the ciphertext is overheard or intercepted because only Z knows his secret decoding procedure. There is no way a curious cryptanalyst, studying Z's public encoding algorithm, can discover Z's decoding algorithm. In principle he might find it, but in practice that would require a supercomputer and a few million years of running time.

An outsider cannot "sign" a message to Z, but any member of the group can. Here is the devilishly clever way the signature works. Suppose A wants to sign a message to Z. He first encodes the plaintext number by using his own secret inverse algorithm. Then he encodes the ciphertext number a second time, using Z's public algorithm. After Z receives the ciphertext, he first transforms it by applying his own secret decoding algorithm, then he applies A's public encoding algorithm. Out comes the message!

Z knows that only A could have sent this doubly encoded ciphertext because it made use of A's secret algorithm. A's "signature" is clearly unforgeable. Z cannot use it to send a message purporting to come from A because Z still does not know A's secret decoding algorithm. Not only that, but if it were to become necessary at some future time to prove to a third party, say a judge in a court of law, that A did in fact send the message, this can be done in a way that neither A, Z nor anyone else can dispute.

Diffie and Hellman suggested in their paper a variety of trapdoor functions that might be used for such systems. None is quite what is desired, but early in 1977 there was a second breakthrough. Ronald L. Rivest, Adi Shamir and Leonard Adleman, computer scientists at the Massachusetts Institute of Technology, developed an elegant way to implement the Diffie-Hellman system by using prime numbers.

Rivest obtained his doctorate in computer science from Stanford University in 1973 and is now an associate professor at M.I.T. Once he had hit on the brilliant idea of using primes for a public cipher system, he and his two collaborators had little difficulty finding a simple way to do it. Their work, supported by grants from the NSF and the Office of Naval Research, appears in *A Method of Obtaining Digital Signatures and Public-Key Cryptosystems* (Technical Memo 82, April 1977), issued by the Laboratory for Computer Science, Massachusetts Institute of Technology, 545 Technology Square, Cambridge, MA 02139.

To explain Rivest's system we need a bit of background in prime-number theory. The fastest-known computer programs for deciding whether a number is prime or composite (the product of primes) are based on a famous theory of Fermat's stating that if $p$ is prime, and $a$ is any positive number less than $p$, then $a^{p-1} = 1$ (modulo $p$). Suppose we want to test a large odd number $n$ (all primes except 2 are of course odd) for primality. A number $a$ is selected at random and raised to the power of $n - 1$, then divided by $n$. If the remainder is not 1, $n$ cannot be prime. For example, $2^{21-1} = 4$ (modulo 21); therefore 21 is composite. What, however, is the connection between 2 (the randomly chosen $a$) and 3 and 7, the two prime factors of 21? There seems to be no connection whatever. For this reason Fermat's test is useless in finding prime factors. It does, however, provide a fast way of proving that a number is composite. Moreover, if an odd number passes the Fermat test with a certain number of random $a$'s, it is almost certainly prime.

This is not the place to go into more details about computer algorithms for testing primality, which are extremely fast, or algorithms for factoring composites, all of which are infuriatingly slow. I content myself with the following facts, provided by Rivest. They dramatize the staggering gap in the required computer time between the two kinds of testing. For example, to test a 130-digit odd number for primality requires at the most (that is, when the number actually is prime) about seven minutes on a PDP-10 computer. The same algorithm takes only 45 seconds to find the first prime after $2^{200}$. (It is a 61-digit number equal to $2^{200} + 235$.)

Contrast this with the difficulty of finding the two prime factors of a 125- or 126-digit number obtained by multiplying two 63-digit primes. If the best algorithm known and the fastest of today's computers were used, Rivest estimates that the running time required would be about 40 quadrillion years! (For a good discussion of computer methods of factoring into primes, see Donald E. Knuth's *Seminumerical Algorithms*, Section 4.5.4.) It is this practical impossibility, in any foreseeable future, of factoring the product of two large primes that makes the M.I.T. public-key cipher system possible.

To explain how the system works, the M.I.T. authors take as an example of plaintext a paraphrase of a remark in Shakespeare's *Julius Caesar* (Act 1, Scene 2): ITS ALL GREEK TO ME.

This is first changed to a single number, using the standard key: A = 01, B = 02, . . . , z = 26, with 00 indicating a space between words. The number is 09201900011212000718050511002015001305.

The entire number is now encoded by raising it to a fixed power $s$, modulo a certain composite number $r$. The composite $r$ is obtained by randomly selecting (using a procedure given in the M.I.T. memorandum) two primes, $p$ and $q$, each of which is at least 40 digits long, and multiplying them together. The number $s$ must be relatively prime to $p - 1$ and $q - 1$. Numbers $s$ and $r$ are made public, to be used in the encoding algorithm. The encoding operation can be done very efficiently even for enormous values of $r$; indeed, it requires less than a second of computer time.

The two prime factors of $r$ are withheld, to play a role in the secret inverse algorithm. This inverse algorithm, used for decoding, consists in raising the ciphertext number to another power $t$, then reducing it modulo $r$. As before, this takes less than a second of computer time. The number *t*, however, can be calculated only by someone who knows *p* and *q*, the two primes that are kept secret.

If the message is too long to be handled as a single number, it can be broken up into two or more blocks and each block can be treated as a separate number. I shall not go into more details. They are a bit technical but are clearly explained in the M.I.T. memo.

To encode ITS ALL GREEK TO ME, the M.I.T. group has chosen *s* = 9007 and  *r* = 1143816257578888676692357799761466120102182967212423 6256256184293570693524573389783059712356395870505898907514 7599290026879543541.

The number *r* is the product of a 64-digit prime *p* and a 65-digit prime *q*, each randomly selected. The encoding algorithm changes the plaintext number (09201 . . . ) to the following ciphertext number: 1999351314978051004523171227402606474232040170583914631037037 1740625971608948927504309920962672582675012893554461353823769748026.

As a challenge to *Scientific American* readers the M.I.T. group has encoded another message, using the same public algorithm. The ciphertext is shown in Figure 92. Its plaintext is an English sentence. It was first changed to a number by the standard method explained above, then the entire number was raised to the 9007th power (modulo *r*) by the shortcut method given in the memorandum. To the first person who decodes this message the M.I.T. group will give $100.

To prove that the offer actually comes from the M.I.T. group, the following signature has been added: 167178611503808442460152713891683982454369010323583112178350384469290626554487922371144905095786086556624965779748400040570 20373.

The signature was encoded by using the secret inverse of the encoding algorithm. Since the reader has no public encoding algorithm of his own, the second encoding operation has been omitted. Any reader who

| 9686 | 9613 | 7546 | 2206 |
|------|------|------|------|
| 1477 | 1409 | 2225 | 4355 |
| 8829 | 0575 | 9991 | 1245 |
| 7431 | 9874 | 6951 | 2093 |
| 0816 | 2982 | 2514 | 5708 |
| 3569 | 3147 | 6622 | 8839 |
| 8962 | 8013 | 3919 | 9055 |
| 1829 | 9451 | 5781 | 5154 |

**Figure 92**
*A ciphertext challenge worth $100*

has access to a computer and the instructions in the M.I.T. memorandum can easily read the signature by applying the M.I.T. group's public encoding algorithm, that is, by raising the above number to the power of 9,007, then reducing it modulo $r$. The result is 0609181 9200019151222051800230914190015140500082114041805040004 15121 2011819. It translates (by the use of the standard key) to FIRST SOLVER WINS ONE HUNDRED DOLLARS. This signed ciphertext could come only from the M.I.T. group because only its members know the inverse algorithm by which it was produced.

Rivest and his associates have no proof that at some future time no one will discover a fast algorithm for factoring composites as large as the $r$ they used or will break their cipher by some other scheme they have not thought of. They consider both possibilities extremely remote. Of course, any cipher system that cannot be proved unbreakable in the absolute sense of one-time pads is open to sophisticated attacks by modern cryptanalysts who are trained mathematicians with powerful computers at their elbow. If the M.I.T. cipher withstands such attacks, as it seems almost certain it will, Poe's dictum will be hard to defend in any form.

Even in the unlikely event that the M.I.T. system is breakable, there are probably all kinds of other trapdoor functions that can provide virtually unbreakable ciphers. Diffie and Hellman are applying for patents on cipher devices based on trapdoor functions they have not yet disclosed. Computers and complexity theory are pushing cryptography into an exciting phase, and one that may be tinged with sadness. All over the world there are clever men and women, some of them geniuses, who have devoted their lives to the mastery of modern cryptanalysis. Since World War II even those government and military ciphers that are not one-time pads have become so difficult to break that the talents of these experts have gradually become less useful. Now these people are standing on trapdoors that are about to spring open and possibly drop them completely from sight.

## ANSWERS

In spite of the many errors in the published version of the cipher Poe could not solve, about a dozen readers, including 16-year-old James H. Andres, were able to crack it. The plaintext is as follows:

MR. ALEXANDER,

HOW IS IT, THAT, THE MESSENGER ARRIVES HERE AT THE SAME TIME WITH THE
SATURDAY COURIER AND OTHER SATURDAY PAPERS WHEN ACCORDING TO THE DATE
IT IS PUBLISHED THREE DAYS PREVIOUS. IS THE FAULT WITH YOU OR THE POSTMASTERS?

The cipher is a polyalphabetic substitution cipher working with 12 alphabets keyed by the words "United States." Each letter indicates the degree of shift for a Caesar cipher. Thus the alphabet key for M, the first letter of the plaintext, is A = U, B = V, C = W and so on. For R, the second letter of the plaintext, the key is A = N, B = O, C = P and so on.

There were 16 errors in the published cryptogram: first, J was given as the third letter instead of I, and second, the fifth letter in the message was omitted. If the second mistake had not been made, Poe might have guessed the opening to be "Mr. Alexander" and the solution would have followed easily.

As for the $100 challenge cipher, no one has cracked it. Rivest told me in 1988 that he no longer has a record of the message or the primes he used. However, since I gave the public key, he will be able to verify a solution if he receives one.

# CHAPTER 14
## Trapdoor Ciphers II

| 9686 | 9613 | 7546 | 2206 |
|------|------|------|------|
| 1477 | 1409 | 2225 | 4355 |
| 8829 | 0575 | 9991 | 1245 |
| 7431 | 9874 | 6951 | 2093 |
| 0816 | 2982 | 2514 | 5708 |
| 3569 | 3147 | 6622 | 8839 |
| 8962 | 8013 | 3919 | 9055 |
| 1829 | 9451 | 5781 | 5154 |

When I wrote the preceding chapter for a column in the August 1977 issue of *Scientific American*, I certainly had not anticipated the intense furor it would arouse. As I reported, Ronald Rivest had offered to send a copy of the M.I.T. memo giving details about what soon came to be known as the RSA cryptosystem (after the initials of the three mathematicians) to anyone who sent M.I.T. a stamped self-addressed envelope. This offer prompted Joseph Meyer, an angry employee of NSA (National Security Agency), to fire off threatening letters to the leaders of a coming symposium on cryptography, warning them that public disclosures of trapdoor systems violated national security laws.

M.I.T. had been flooded with some 7,000 requests from all over the world for its trapdoor-cipher memo, but Meyers's letter put a stop to the mailing. It was almost a year before M.I.T. attorneys concluded that the memo violated no laws and allowed the mailing to be resumed. Since then an uneasy truce has prevailed between NSA and researchers on public-key cryptosystems. There has been no outright censorship and no one has gone to jail, but there has been much voluntary censorship by mathematicians. High-level research within NSA remains top secret, and it is impossible for outsiders such as myself to know what NSA knows. The acronym NSA, it has often been said, stands for "Never Say Anything" or (reflecting NSA's efforts to keep out of the limelight) "No Such Agency."

It is not hard to understand why NSA became so jittery. Publication of seemingly break-proof ciphers obviously allows other nations to adopt codes that NSA might be unable to crack, and, as long as there is freedom here to publish techniques for breaking such ciphers, any nation using a breakable code would at once stop using it. Moreover, as I implied, if truly unbreakable codes become common around the world, it would almost put NSA out of business.*

For many years U.S. banks and corporations have been protecting their communications with a system called Data-Encryption Standards (DES), developed by IBM and approved by the National Bureau of Standards. The DES is a "symmetric" system, meaning that it codes and decodes by the same procedure, not an "asymmetric" trapdoor system. Nevertheless, it is extremely difficult to break if its key uses a large number of bits. There is evidence that NSA persuaded IBM to hold its key size down to 56 bits so that in case foreign governments chose to adopt DES, NSA could still break their codes. Although DES is still being used, Bell Telephone rejected it for security reasons, and it has come under heavy fire, especially from Diffie and Hellman, who consider it too weak to survive many more years.

Chief rivals to the RSA system have been the so-called knapsack systems. Knapsack problems are a large family of combinatorial tasks that involve finding among a set of numbers a subset that will "fit," subject to various constraints, inside a hypothetical "knapsack." The simplest example, known as the subset-sum problem, is to select from a set of integers a subset that will add to a specified value. Subset-sum problems are common in the puzzle books of Sam Loyd and Henry Dudeney, often in the form of a target whose concentric rings are assigned different numerical values. The task is to determine how shots can be fired at the target to obtain hits that add exactly to a given sum. Such puzzles are not hard to solve by trial and error when the set of numbers is small, but they become enormously difficult as the set increases in size.

A combinatorial task is called "hard" if it can be shown that no computer algorithm can solve it in "polynomial time." This results from

---

*A good discussion of the pros and cons of the debate between NSA's desire for security and the mathematical community's desire for openness will be found in David Kahn's 1983 book, *Kahn on Codes*, pp. 198–203.

the fact that as a certain parameter of the problem increases, the time required to solve the problem grows at an exponential, or "nonpolynomial," rate. Studies of such problems belong to a new branch of mathematics and computer science called "complexity theory." A great deal of work has been done and is continuing on a special class of problems called NP-complete (NP for nondeterministic-polynomial). There are now hundreds of such problems, all believed to be hard (though no proof has yet been found), and all related so that if an algorithm is found for solving one of them in polynomial time, it will at once solve all of them. The subset-sum problem is NP-complete.

Ralph Merkle was the first to base a knapsack system on subset-sum, and for a short time it was preferred to RSA because it was faster to code and decode. Then in 1982 Adi Shamir, the Israeli member of the M.I.T. team, found an algorithm that solved "almost all" knapsack systems in polynomial time. Ralph Merkle had offered a $100 prize to anyone breaking his system, and Shamir collected it. Merkle then increased the complexity of his system to what he called a "multiply iterated" version, offering $1,000 to anyone who could break it. Ernest Brickell of Sandia Laboratories won the second prize in 1984. The subset-sum problem continues, however, to be NP-complete, and it is possible that new cipher systems based on it or on other knapsack problems will withstand the onslaught of new algorithms. Rivest and B. Chor have proposed a knapsack system based on the logarithms of large primes that has not so far been cracked by the Sandia techniques. It has been reported that NSA thought of knapsack codes about a decade before Merkle did but, in keeping with its "Never Say Anything" policy, has kept mum about it.

The factoring of large numbers is not in the NP-complete family, but it is thought to be hard, and so far no one has found a way to factor large numbers in polynomial time. However, such techniques have been steadily improving along with methods of testing the primality of big numbers. Fast procedures for testing primality in "near polynomial time" were discovered in the 1980's, and in 1982 a team at Sandia Labs, under the direction of Gustavus Simmons, succeeded (with a Cray supercomputer) in factoring the Mersenne number $2^{521} - 1$, an integer of 157 digits. It took the Cray about 32 hours to find the number's three prime factors. Until this breakthrough, mathematicians had estimated that a Cray computer would need millions of years to factor a number with more than 100 digits.

In view of these new factoring techniques, no one can rule out the possibility of a polynomial-time algorithm that would topple the RSA system. When the system was first announced, numbers of 80 digits were recommended for the two primes $p$ and $q$. It is now recommended that each of these primes be at least 100 digits. It is best that they be nearly the same size, that $p$ plus or minus 1 and $q$ plus or minus 1 should each have at least one large prime factor and that the greatest common divisor of $p - 1$ and $q - 1$ be fairly small. Up to now the RSA system has remained secure. Computer chips for fast coding and decoding are available from RSA Data Security, Inc., 10 Twin Dolphin Drive, Redwood City, CA 94065.

A variety of fascinating spin-offs have resulted from the basic ideas behind trapdoor codes. It occurred at once to Robert Floyd, a computer scientist at Stanford, that such systems could be used by two people in communication by mail (paper or electronic) to make random decisions in ways that are immune to cheating. For example, two people in touch by telephone can agree on the outcome of a random flip of a coin or the outcome of a die toss. In June 1978 Floyd sent me a letter outlining how two persons could play backgammon by mail or telephone. Picking up Floyd's cue, Rivest, Shamir and Adleman wrote a paper on "mental poker" in which they explained how two players who did not trust each other can actually play a fair game of poker over the phone without using any cards.*

Another spin-off was the development of ingenious systems for making secure the transmission of scientific data over electronic networks. Consider, for instance, research conducted by instruments that have been landed on Mars. Researchers need to be sure that when they link to these instruments, they are not linked to some other data source and that no one else can alter the data being transmitted or can alter their instructions to the instruments. In brief, they need to be assured of the network's authenticity, integrity and secrecy.†

The most startling, almost unbelievable, spin-off has been the development of what are called "zero-knowledge proofs." Suppose a mathematician discovers a proof of a certain theorem. He wants to convince his colleagues that he actually has the proof but doesn't want to disclose the proof itself. In 1986 it was shown that this could be done with special

---

*"Mental Poker" was first published in 1979 as a technical report of the M.I.T. Laboratory of Computer Science. It is reprinted in *The Mathematical Gardner*, edited by David Klarner (Prindle, Weber, and Schmidt, 1981). On coin flipping, see "Coin Flipping by Telephone," Manuel Blum, *SIGACT News*, 15, 1983, pp. 23–27.

†For a good summary of recent developments in this area of "telescience," see Peter Denning's "Security of Data in Networks," in *American Scientist*, 75, January/February 1987, pp. 12–14. For more detailed information see Dorothy Denning's book *Cryptography and Data Security* (Addison-Wesley, 1982).

cases of NP-complete problems. For example, consider the NP-complete task of finding a Hamiltonian circuit—a path that goes through all points of a graph just once and returns to the starting point. Suppose that for a given graph with a large number of points it is not known whether it has a Hamiltonian circuit. A mathematician wishes to convince his colleague that he has found such a circuit, but he doesn't want to reveal the actual circuit. It is hard to comprehend, but there are now techniques by which he can do this.

In 1986 Manuel Blum, a computer expert at the University of California, Berkeley, found a way to apply zero-knowledge proofs to *any* mathematical problem! The procedure consists essentially of a dialogue between the "prover" and the "verifier" who wants to be convinced that the proof exists. The verifier asks a series of random questions, each to be answered by yes or no. After the first question, the verifier is convinced that the prover has a 1/2 chance of being wrong. After the second question, he is convinced the prover has a 1/4 chance of being wrong. After the third question, the probability drops to 1/8, and so on, with the denominators increasing in a doubling series. After, say, 100 questions the chance that the prover is lying or doesn't have a proof becomes so close to zero that the verifier is convinced beyond any shadow of a doubt. After 300 questions the denominator is $2^{300}$, which is more than the number of atoms in the universe. There is never absolute certainty that the proof exists, but it is so close to certainty that all doubt vanishes. See the Bibliography for some nontechnical pieces about this surprising new development.

Do zero-knowledge proofs have practical applications beyond satisfying the egos of mathematicians who want to announce a discovery before anyone else does and before they have published the details? They do indeed. Adi Shamir, now at the Weizman Institute in Israel, found a way to make use of zero-knowledge methods for creating unforgeable ID cards. Think of a computer chip within such a card that can engage in rapid dialogue with a computer chip in an instrument used for verifying the ID. Within a few seconds enough random questions have been asked and answered to convince the verifier "beyond any shadow of a doubt," even though the verification cannot be absolutely certain. There have for decades been methods of showing that large numbers are almost certainly prime, such as by using probabilistic techniques, but finding similar methods for validating an ID card came as a big surprise.

The implications of unforgeable ID cards for military as well as civilian use are so enormous that when Shamir applied for a U.S. patent, the Army ordered that all documents and materials related to such cards be destroyed. This aroused such a storm of protest from the mathematical community that the government quickly rescinded the order, giving as its reason that it could not impose such restraints on a mathematician who was not a U.S. citizen. No one knows if NSA had any role in this attempt at censorship. For a good account of the flap, see the *New York Times* 1987 article cited in the Bibliography.

Inventions of new public-key cryptosystems and new ways to break them, as well as their applications to the security of networks and to identification techniques, are occurring so rapidly that by the time you read this chapter, much of it may be out of date. The science of cryptology is undergoing a curious revolution, and no one can predict just where it will lead. Let me close with some whimsical dialogue from *Romanoff and Juliet*, a play by Peter Ustinov first produced in New York City in 1957.

The scene occurs at the close of the second act. The General (played by Ustinov) is president of what is identified only as the smallest nation in Europe. Hooper Molesworth is the country's American ambassador. Vadim Romanoff is the Russian ambassador. Molesworth's daughter Juliet and Romanoff's son Igor are lovers.

In the American Embassy, the General says to Molesworth:

"Incidentally, they [the Russians] know your code."

Molesworth replies: "We know they know our code. We only give them things we want them to know."

Crossing over to the Russian Embassy, the General remarks to Romanoff:

"Incidentally, they [the Americans] know you know their code."

Romanoff says: "That does not surprise me in the least. We have known for some time that they knew we knew their code. We have acted accordingly—by pretending to be duped."

Returning to the American Embassy, the General says to Molesworth:

"Incidentally, you know—they know you know they know you know. . . . "

Molesworth is now genuinely alarmed.

"What? Are you sure?"

"I'm positive."

"Thank you — thank you! I shan't forget this."

The General is amazed.

"You mean you didn't know?"

"No!"

In 1957 a dialogue like this was at least believable. Could it occur today? Maybe NSA knows.

## BIBLIOGRAPHY

*Books on cryptanalysis*

*Cryptanalysis*. Helen Gaines. Dover, 1956. Reprint of a 1939 book.
*The Codebreakers*. David Kahn. Macmillan, 1967. Supersedes all previous histories.
*Elementary Cryptanalysis*. A. Sinkov. Random House, 1968.
*Cryptography: A Primer*. Alan Konheim. Wiley, 1981.
*Codes, Ciphers, and Computers*. Bruce Bosworth. Hayden, 1982.
*Cryptography: A New Dimension in Computer Data Security*. C. H. Meyer and S. M. Matyas. Wiley, 1982.
*Kahn on Codes: New Secrets of Cryptology*. David Kahn. Macmillan, 1983.
*Codes, Ciphers, and Secret Writing*. Martin Gardner. Dover, 1984. Reprint of a 1972 book for children.
*Cryptographic Significance of the Knapsack Problem*. Luke J. O'Connor and Jennifer Seberry. Aegean Park Press, 1988.

*On Poe and cryptanalysis*

"A Few Words on Secret Writing." Edgar Allan Poe, in *Graham's Magazine*, 19, July 1841, pp. 33–38.
"Edgar Allan Poe, Cryptographer." William Friedman, in *American Literature*, 8, November 1936, pp. 226–280.
"What Poe Knew About Cryptography." William Wimsatt, Jr., in *Publications of the Modern Language Association*, 58, 1945, pp. 754–779.
"Poe's Challenge Cipher Finally Broken." Brian Winkel, in *Cryptologia*, 1, January 1977, pp. 93–96. For solution and comments on how they were obtained see the same journal, 1, October 1977, pp. 318–325.

## News reports

"Cryptic Reaction." Richard Shaffer, in *The Wall Street Journal*, June 16, 1978, p. 1.

"An Uncrackable Code?" *Time*, July 3, 1978, pp. 55–56.

"Opening the 'Trapdoor Knapsack.'" Philip Faflick, in *Time*, October 28, 1982.

Articles by Gina Bara Kolata in *Science*: "Computer Encryption and the National Security Agency Connection," July 29, 1977, pp. 438–448; "Cryptography: On the Brink of a Revolution?," August 19, 1977, pp. 747–748; "Cryptology: Scientists Puzzle Over Threat to Open Research, Publication," September 30, 1977, pp. 1345–1349; "Cryptology: A Secret Meeting at IDA," April 14, 1978, p. 184; "New Codes Coming Into Use," May 16, 1980, pp. 694–695; "Prior Restraints on Cryptography Considered," June 27, 1980, pp. 1442–1443; "Testing for Primes Gets Easier," September 26, 1980, pp. 1503–1504; "Cryptographers Gather to Discuss Research," November 1981, pp. 646–647; "Another Promising Code Falls," December 16, 1983, p. 1224.

## Nontechnical articles

"The New Unbreakable Codes: Will They Put NSA Out of Business?" Deborah Shapley, in *The Washington Post Outlook*, Section B1, July 9, 1978.

"The Mathematics of Public-Key Cryptography." Martin Hellman, in *Scientific American*, August 1979, pp. 146–157.

"Unbreakable Code." Roger Rapoport, in *Omni*, September 1980, pp. 84–86, 92.

"Safety in Numbers." George Davida, in *The Sciences*, July/August 1981, pp. 9–14.

"Cryptology: From Caesar Ciphers to Public-Key Cryptosystems." Dennis Luciano and Gordon Prichett, in *The College Mathematics Journal*, 18, January 1987, pp. 2–17.

## Advanced articles

"New Directions in Cryptography." Whitfield Diffie and Martin Hellman, in *IEEE Transactions of Information Theory*, November 1976, pp. 644–654.

"A Method of Obtaining Digital Signatures and Public-Key Cryptosystems." Ronald Rivest, Adi Shamir and Leonard Adleman, M.I.T. Laboratory for Computer Science, Technical Memo 82, April 1977. Reprinted in *Communications of the ACM*, 21, February 1978, pp. 120–126.

"Preliminary Comments on the M.I.T. Public-Key Cryptosystem." Gustavus Simmons and Michael Norris, in *Cryptologia*, 1, October 1977, pp. 406–414. See Rivest's reply in the same journal, 2, January 1978, pp. 62–65.

"Secure Communications Over Insecure Channels." R. C. Merkle, in *Communications of the ACM*, 21, April 1978, pp. 294–299.

"Hiding Information and Signatures in Trapdoor Knapsacks." R. Merkle and M. Hellman, in *Transactions of Information Theory*, September 1978, pp. 525–530.

"Cryptology: The Mathematics of Secure Communication." G. J. Simmons, in *Mathematical Intelligencer*, 1, 1979, pp. 233–246.

"Privacy and Authentication: An Introduction to Cryptography." Whitfield Diffie and Martin Hellman, in *Proceedings of the IEEE*, 67, March 1979, pp. 397–427.

"Secrecy, Authentication, and Public-Key Systems." Ralph Charles Merkle, Technical Report 1979-1, Information Systems Laboratory, Stanford University, June 1979.

"Symmetric and Asymmetric Encryption." Gustavus Simmons, in *Computing Surveys*, 11, December 1979, pp. 305–330. Bibliography has 66 references.

"Error-Correcting Codes and Cryptography." N. J. A. Sloane, in *The Mathematical Gardner*. Prindle, Weber, and Schmidt, 1981.

"A Polynomial Time Algorithm Testing for Breaking the Basic Merkle-Hellman Cryptosystems." A. Shamir, in *Proceedings of the 23rd Annual Symposium of the Foundations of Computer Science*, 1982, pp. 145–152.

"On Breaking the Generalized Knapsack Public-Key Cryptosystem." L. M. Adleman, in *Proceedings of the 15th ACM Symposium on Theory of Computing*, 1983, pp. 402–412.

"How to Exchange (Secret) Keys." Manuel Blum, in *ACM Transactions on Computer Systems*, 1, May 1983, pp. 175–193.

"Integer Programming and Cryptography." H. W. Lenstra, Jr., in *Mathematical Intelligencer*, 6, 1984, pp. 14–19.

"Proof Checking the RSA Public-Key Encryption Algorithm." Robert Boyer and J. Strother Moore, in *The American Mathematical Monthly*, 91, March 1984, pp. 181–189.

"Cryptology." Ronald Rivest, in the *Handbook of Theoretical Computer Science*, Chapter 13, 1988. An excellent overview of the current revolution in cryptography. The bibliography runs to more than 170 references.

### On the DES system

"Assessment of the National Bureau of Standards Proposed Federal Data Encryption Standard." Robert Morris, N. J. A. Sloane and A. D. Wyner, in *Cryptologia*, 1, 1977, pp. 281–306.

"Exhaustive Cryptanalysis of the NBS Data Encryption Standard." W. Diffie and M. E. Hellman, in *Computer*, 10, June 1977, pp. 74–84.

"DES Will Be Totally Insecure Within Ten Years." M. E. Hellman, in *IEEE Spectrum*, 16, July 1979, pp. 32–39.

### On zero-knowledge proofs

"Keeping Secrets: How to Prove a Theorem So That No One Else Can Claim It." Ivars Peterson, in *Science News*, 130, August 30, 1986, pp. 140–141.

"Zero-Knowledge Proofs." Joe Buhler, in *Focus* (newsletter of the American Mathematical Association), 6, October 1986, pp. 1, 6–7.

"A New Approach to Protecting Secrets is Discovered." James Glieck, in the *New York Times*, February 17, 1987, pp. 17–18.

"Brief U.S. Suppression of Proof Stirs Anger." *New York Times*, February 17,

*Contemporary Cryptography.* Gustavus Simmons. IEEE Press, 1992.

*Public-Key Cryptography.* T. Beth, M. Frish, and G. J. Simmons (eds.) Springer-Verlag, 1992.

"Cipher Probe." William Bulkeley, in *The Wall Street Journal*, April 28, 1994, page 1ff.

"Suddenly, Number Theory Makes Sense to Industry." Fred Guterl, in *Business Week*, June 20, 1994, pp. 172–74.

"Lost in Kafka Territory." *U. S. News*, April 3, 1995, pp. 32–33.

"The Encryption Wars." Steven Levy, in *Newsweek*, April 24, 1995, pp. 55–56.

"Whitfield Diffie," an interview, in *Omni*, Winter 1995, pp. 87–93.

"Confidential Communication on the Internet." Thomas Beth, in *Scientific American*, December 1995, pp. 88–91.

# CHAPTER 15

# Hyperbolas

Lewis Carroll once wrote (in *The Dynamics of a Parti-cle*): "What mathematician has ever pondered over an hyperbola, mangling the unfortunate curve with lines of intersection here and there, in his efforts to prove some property that perhaps after all is a mere calumny, who has not fancied at last that the ill-used locus was spreading out its asymptotes as a silent rebuke, or winking one focus at him in contemptuous pity?"

Put a large sphere, say a basketball, on a light-colored tabletop in a darkened room. Shine a flashlight directly down on the ball as is shown at A in Figure 93. The ball's shadow is, of course, a circle. Its center is the point where the ball touches the surface of the table.

Move the flashlight toward the east as at B in the illustration. The shadow stretches to an ellipse. The center of the circle has now split into two points that are the foci of the ellipse. The sphere rests on the focus closer to the light source. As you move the light farther east, the other focus moves west, increasing the eccentricity of the ellipse.

Lower the light source until it is level with the top of the ball (C). The ball still rests on the eastern focus, but now the western focus has traveled to infinity. The outline of the shadow is a parabola.

Move the light until it is below the top of the ball (D). The curve

**205**

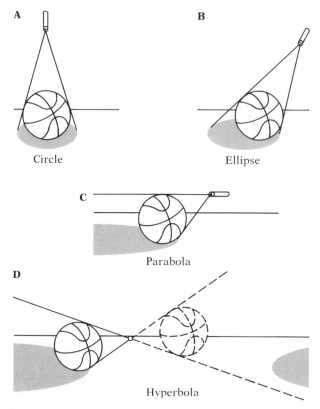

**Figure 93** *Shadow experiment displays four conicsection curves*

of the shadow becomes a hyperbola. The sphere still touches the table at the focus of the hyperbola but now something delightful has happened to the missing focus.

Imagine that in all four pictures there is a counterball, identical with the sphere on the table but placed on the other side of the light source the same distance away. The counterball is shown with a broken line in the last picture (D). Note that it casts a shadow cone identical with the one produced by the ball on the table but turned the other way. The apexes of the cones meet at the light.

In the first three pictures the counterball's shadow lies above the plane of the tabletop. When the light source moves below the top of the ball, however, the countershadow falls on the plane to form a counter-curve that is the eastern branch of the hyperbola under the ball. The missing focus has, so to speak, traveled around infinity to return on the other side! Since the two ends of infinity are one and the same point, they are comparable to the extremities of a ring that has been cut and opened out into a straight line. This infinite, unmutilated ring is the geometric metaphor behind Henry Vaughan's famous couplet "I saw Eternity the other night / Like a great *Ring* of pure and endless light."

Imagine the counterball in the last picture moving away from the light source but continuously expanding so that it always touches the sides of the countercone. When the counterball is large enough to touch the tabletop, it will rest on the focus of the counterbranch of the hyperbola. These two spheres of unequal size, nested in the cones and touching the cutting plane at the foci of the hyperbola, provide an old, elegant proof that the curves are indeed branches of a hyperbola. The proof is clearly explained on pages 8 and 9 of *Geometry and the Imagination*, by David Hilbert and S. Cohn-Vossen. If both spheres are placed in the same cone, a similar proof can be used (see Chapter 15 of my *New Mathematical Diversions from Scientific American*) to show that a curve is an ellipse.

The shadow experiment I have been describing is a way of displaying the four curves as conic sections. The plane of the tabletop is the plane that cuts the cones. It is apparent that the circle is a limiting case of the ellipse. The parabola is a limit of both ellipse and hyperbola. Like the circle, it has only one shape, although the shape can be enlarged or diminished. Both the ellipse and the hyperbola, however, are infinite families of different shapes.

Astronomers often find it difficult to decide whether a comet or a meteor is traveling an elliptical, parabolic or hyperbolic path. It is easy to see why. Vary the parabola ever so slightly one way and it becomes an ellipse. Vary it ever so slightly the other way and it becomes a hyperbola. Comets in permanent orbit around the sun move on ellipses. Those that enter the inner solar system and then leave, never to return, move on parabolas or hyperbolas.

Since the hyperbola is a kind of ellipse split in half by infinity, it is not surprising that the two curves are related to each other in many inverse ways. An ellipse is the locus of all points whose distances from two fixed points have a constant sum. The two fixed points are called the foci of the curve. This is the basis for the ancient method of drawing an ellipse with a pencil and a loop of string that goes around two pins.

The hyperbola is the locus of all points whose distances from two fixed points have a constant difference. Figure 94 shows a string device for drawing one branch of a hyperbola. The pencil at P keeps the string taut and pressed against the rod as the rod rotates around its fixed end at focus *A*. The string is attached to focus *B* and to the rod's free end, *C*.

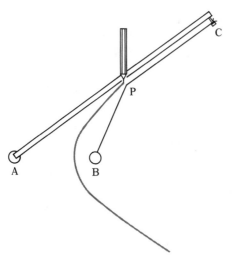

**Figure 94**
*String method of drawing a hyperbola*

$BP + PC$ is constant; therefore $AP - BP$ must also be constant. Since $AP$ and $BP$ are the distances of $P$ from the two foci, we have proved that the curve must be a hyperbola.

Ellipses and hyperbolas are also easily produced by a paper-folding method that points up the inverse kinship of the curves. Draw a circle on a sheet of translucent paper such as waxed paper or tracing paper. Mark a spot anywhere inside the circle that is not at the circle's center. Fold the paper in a variety of ways each of which brings the spot onto the circumference of the circle. Each fold line is tangent to an ellipse. When enough folds have been made, the ellipse will take shape as the "envelope" of the tangents. The spot and the center of the circle are the two foci of the ellipse. To fold a hyperbola follow the same procedure with a spot anywhere *outside* rather than inside the circle. Both branches of the hyperbola appear in Figure 95. Again the spot and the center of the circle are the two foci of the curve.

Will a spot on the circumference of the circle give rise to a parabola? Unfortunately it will not. We can blame this perversity on the parabola's lost focus. Each fold line goes through the center of the circle and is therefore tangent to a degenerate ellipse whose length is the circle's radius and whose width is zero. To get a parabola we need a circle expanded to infinity so that its circumference is a straight line. Rule a line on the paper and pick a point off the line; now the same folding technique will produce a splendid parabola. The missing focus is the "center" of the infinite circle.

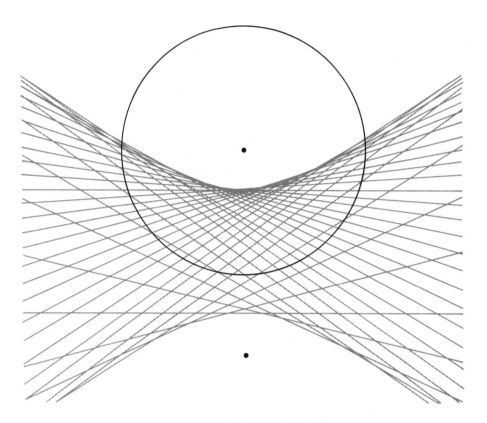

**Figure 95** *Folding a hyperbola*

Of the four conic curves, the hyperbola is the curve least often observed in everyday life. Circles and ellipses are everywhere. We see parabolas whenever we water a lawn with a hose or watch the flight of a baseball.* One of the few times we see a complete hyperbola is when a lamp with a cylindrical or conical shade, open at both ends, throws shadows on a nearby wall. Our forebears saw a branch of a hyperbola on a wall when they held near the wall a candle burning in a candlestick with a circular base.

Scientists and mathematicians are constantly seeing hyperbolas as the graphs of various second-degree equations. Even the simple equation $ab = c$, where $c$ is a constant, graphs as a hyperbola. It is the equation of

---

*Many readers quite properly chided me for saying that tossed objects follow paths that are parabolas. The paths are very close to parabolas, but strictly speaking (and ignoring air resistance) a tossed object follows an elliptical orbit around the earth's center of gravity.

hundreds of physical laws (Boyle's law and Ohm's law, to mention two) and also the equation of many economic functions. A simple experiment to display $ab = c$ can be made with two rectangular sheets of glass. Place them together at one pair of edges and separate the opposite edges by a tiny distance with a short strip of cardboard or a pair of matches. Rubber bands will keep the plates fixed in this position. Stand the device in colored water. Capillary action will make the hyperbola shown in Figure 96.

A typical hyperbola is shown in Figure 97. The two gray lines are the curve's asymptotes, or the unreachable limits the curve's branches approach as they are extended. If the asymptotes, are perpendicular to each other (they are not here), the hyperbola is called equiangular or rectangular.

The arms of a parabola quickly become almost indistinguishable from parallel lines. In contrast, the arms of a hyperbola rapidly widen on their way to infinity, although they are forever confined within the angle of their asymptotes. It is this beautiful property that has inspired many poetic and theological metaphors. Miguel de Unamuno, the Spanish philosopher, called the hyperbola a tragic curve. "I believe that if the geometrician were to be conscious of this hopeless and desperate striving of the hyperbola to unite with its asymptotes," Unamuno wrote, "he would represent the hyperbola to us as a living being and a tragic one!"

In his essay "The Immortality of the Soul," however, Joseph Addison regarded the metaphor with optimism. After death the soul moves ever closer to God without ever becoming God. "We know not yet what we shall be, nor will it ever enter into the heart of man to conceive the glory that will be always in reserve for him. The soul, considered with its

**Figure 96**
*Hyperbola produced by capillary action*

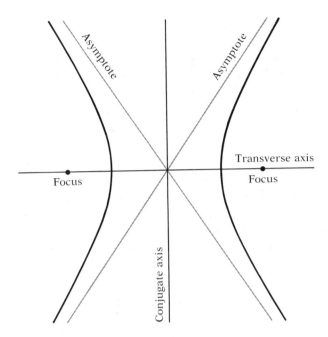

**Figure 97**  *A typical hyperbola*

Creator, is like one of those mathematical lines that may draw nearer to another for all eternity without a possibility of touching it."

Hyperbolas have a dramatic application in range-finding. To understand how this works consider a person *A* who fires a rifle at a distant gong *B*. Assuming that the ground is flat, where must one stand to hear the sound of the gun and the sound of the gong simultaneously?

Let *x* be the distance sound travels in the length of time it takes the bullet to travel from the gun to the gong. *A* and *B* are the foci of countless hyperbolas. The person who hears the sounds simultaneously must stand on a branch of a hyperbola (the one nearest the target) that is the locus of all points the difference of whose distances from *A* and *B* is *x*.

A distant sound can be located by two pairs of listening posts: *A* and *B*, *C* and *D*. Listeners at *A* and *B* record the time they hear the sound. Their clocks are synchronized so that they can obtain a precise difference between the two times. Call the difference *x*. The sound must come from a branch of a hyperbola (the one nearest the sound) that is the locus of all points the difference of whose distances from *A* and *B* is *x*. This curve is drawn on a map. Listeners at *C* and *D* do the same thing and plot a branch of another hyperbola on the same map. The spot where the two curves intersect, nearer the sound's origin, gives the "fix" of the origin.

Hyperbolic navigation systems such as loran, which was developed during World War II, operate by a reverse procedure. Somewhere on the shore a pair of stations $A$ and $B$—one is called the master station, the other the slave station—send out simultaneous radio signals. Another pair of master and slave stations, $C$ and $D$, do the same thing from another shore position. A ship or an airplane at sea, using the time differences in receiving the signals from both pairs of stations, can plot two hyperbolas that intersect on the map to fix its location.

Mirrors with a hyperbolic cross section are found (usually with other kinds of mirrors) in some reflecting telescopes and special-purpose cameras and behind the light sources of flashlights and searchlights. If a light source is at one focus of an elliptical mirror, all the reflected rays converge on the other focus. If the light source is at the focus of a parabola, the reflected rays are parallel as they seek the lost focus at infinity. A hyperbolic mirror causes the reflected rays to diverge as is shown in Figure 98. If, however, we extend these diverging rays backward as is shown by the broken lines, they obligingly converge on the

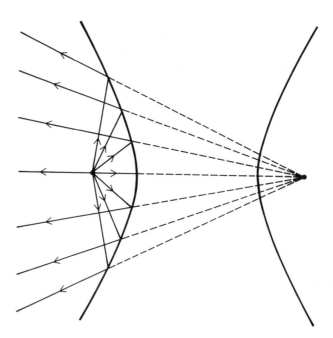

**Figure 98**  *Reflected rays from light at one focus of a hyperbola find the other focus*

**Figure 99** *Cylindrical string model* (left) *twists into a hyperboloid* (middle) *and a double cone* (right)

other focus. In a sense they have traversed the infinite and found the lost focus at a point just behind their origin.

A lovely hyperbolic surface with many remarkable properties is the "hyperboloid of one sheet," first described by Archimedes. A string model of this surface is shown in the middle of Figure 99. Vertical cross sections of the surface are hyperbolas and horizontal cross sections are ellipses. If the horizontal sections are circles, it is a "hyperboloid of revolution of one sheet," so called because it is generated by spinning a hyperbola around its conjugate axis. (If a hyperbola is rotated around its transverse axis, it generates a hyperboloid of rotation of two sheets: a pair of domelike structures separated by a finite distance.)

In 1669 Christopher Wren, the architect who designed St. Paul's Cathedral, reported an extraordinary discovery about the hyperboloid of one sheet. He showed that the hyperboloid of one sheet is what mathematicians now call a ruled surface, a surface consisting of an uncountable infinity of perfectly straight lines!

A cylinder, for example, is a ruled surface of parallel straight lines. A cone is a ruled surface of straight lines that meet at the apex of the cone. The hyperboloid is a ruled surface of two distinct families of straight lines. In the middle of Figure 99 you see some members of one set all slanting the same way with no two lines intersecting. The other family (which is not shown) is a mirror-image set, slanting the other way. Each line of one set, when extended, intersects each extended line of the other. A straight line from each set passes through every point on the hyperboloid. The pair of lines through a point defines the plane tangent to the surface at that point.

It is easy to see from the string model that a hyperboloid of one sheet is generated by the rotation around an axis of a straight line segment that is skew to the axis; that is, the line incorporating the segment does not intersect the axis. (If the rotating line segment is parallel to the axis, it generates a cylinder; if its extension intersects the axis, it generates part of a cone.) This suggests a simple experiment with a pencil and a paper clip. Open the paper clip to form an acute angle and then push one end through the eraser of the pencil, as shown in Figure 100. Turn the wire so that *AB*, the upright part, is skew to the vertical axis of the pencil. Place the pencil between your palms and spin it by sliding your hands rapidly back and forth. With the right lighting, the rotating skew line will form a transparent hyperboloid.

If a cube is spun on one corner, its six skew lines generate a similar surface. With a little practice you can snap a die between your finger and your thumb and make it spin on a corner. Lower your head to view the spinning cube from the side. You will see in profile a hyperboloid between two cones, as shown in Figure 101.

Making the string model of Figure 99 is not difficult. Simply thread a string back and forth through holes evenly spaced around the rim of two cardboard or plywood disks. Spots of glue placed over the holes will keep the cord from sliding. When the disks are held apart to stretch the cord vertically, as they are at the left in the illustration, they model a cylinder. Twist one of the disks 180 degrees clockwise and the strings

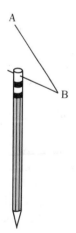

*Figure 100*  *How to spin a hyperboloid*

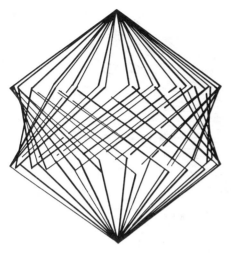

**Figure 101**
*Hyperboloid between cones from spinning cube*

model two cones as shown at the right. Between these two limits, twists produce an infinite family of hyperboloids such as the one shown in the middle of the illustration, ruled with straight strings that slant one way. Twist the disk counterclockwise and you can run through the mirror-image set of hyperboloids, ruled by strings slanting the other way.

A much more difficult model to make, built of rigid wires instead of strings, is described in *Geometry and the Imagination*, pages 16 and 17. The pair of wires at each intersection are joined by a universal joint that allows rotation but not sliding. You would expect such a structure to be rigid. Instead it is flexible in a curious way. If the model is compressed in one direction, the elliptical cross section degenerates into a straight line and the rods fold up into a vertical plane on which they form the envelope of a hyperbola. If the model is collapsed the other way, the rods fold down into a horizontal plane on which they form the envelope of an ellipse. Figure 102, based on a photograph in *Geometry and the Imagination*, shows how two hyperboloids of revolution can provide a gear transmission between two skew axes. The cogs of each gear are one of its sets of generating straight lines. This is only one of many ways that hyperboloids have been used in mechanical linkages.

A striking architectural use of a hyperboloid of revolution of one sheet is provided by the McDonnell Planetarium at Forest Park in St. Louis (see Figure 103). The designer, Gyo Obata, chose the surface because the hyperbolic paths of certain comets suggest, as he put it, "the drama and excitement of space exploration." Note the straight line of the shadow thrown by the circular roof as sunshine slants down on the

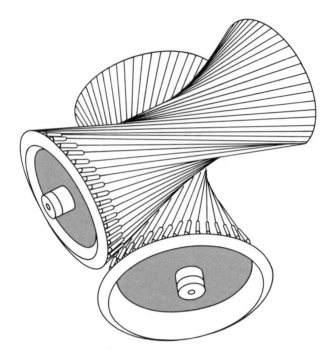

**Figure 102**  *Hyperboloidal gears transmit motion to a skew shaft*

**Figure 103**  *The McDonnell Planetarium in St. Louis, Mo.*

planetarium. Is this shadow line one of the generating straight lines of the surface, or is it a space curve that appears straight only when viewed from the angle shown?

## ANSWERS

Does the straight-line shadow on the photograph of the McDonnell Planetarium in St. Louis corresponds to one of the straight lines that rule the surface of the structure's hyperboloid of revolution of one sheet? The answer is yes. At those times when the line containing the edge of the shadow intersects the sun, the edge of the shadow on the side of the planetarium coincides with a generating line of the surface. That line is a projection of a single point on the circumference of the circular top of the planetarium, not a projection of the entire circumference. When the sun is at any other elevation, the edge of the shadow will be a curved line and not a generating line of the surface.

## ADDENDUM

Pierre Bézier wrote from Paris to tell me that hyperboloids of one sheet are often used for the construction of cooling towers at power stations. Because the concrete sides can be reinforced by straight steel bars (as in the model explained earlier), the structure is unusually sturdy. The conning towers of old battleships, Clyde Holvenstot informed me, were also often built the same way. He sent me a photograph from the *New York Times Book Review* (October 2, 1977) showing two such hyperboloid towers on the battleship *USS Michigan* before the Navy scrapped her.

Derek Ball, in a 1980 paper (see the Bibliography), discusses an unexpected appearance of hyperbolas in constructions involving lines that cut the area of an equilateral triangle into integral fractions. For example, if you draw across the triangle all straight lines that exactly bisect its area, the envelope of the lines is the small delta-shaped figure shown in Figure 104. Its sides are hyperbolas.

In my remarks about how to fold a circle to produce ellipses and hyperbolas and why such folding will not produce a parabola, I failed to acknowledge letters from Walter Cibulskis of the Illinois Institute of Technology, who called these facts to my attention.

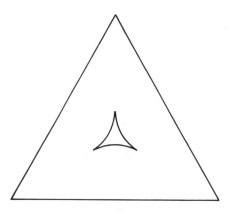

**Figure 104**
*Three hyperbolas form the envelope
of bisecting lines*

## BIBLIOGRAPHY

"Conic Sections." Henry Martin Taylor, in *Encyclopedia Britannica*, Ninth Edition. Encyclopedia Britannica, Inc., 1890.

*Geometry and the Imagination*. David Hilbert and S. Cohn-Vossen. Chelsea, 1952.

*A Book of Curves*. E. H. Lockwood. Cambridge University Press, 1961.

"Halving Envelopes." Derek Ball, in *Mathematical Gazette*, 64, October 1980, pp. 166–172.

# CHAPTER 16

# The New Eleusis

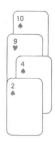

> I shall always consider the best guesser the best
> Prophet.
>
> —CICERO *De Divinatione*

> Don't never prophesy—onless ye know.
>
> —JAMES RUSSELL LOWELL, *The Biglow Papers*

In June 1959, I had the privilege of introducing in *Scientific American* a remarkable simulation game called Eleusis. The game, which is played with an ordinary deck of cards, is named for the ancient Eleusinian mysteries, religious rites in which initiates learned a cult's secret rules. Hundreds of ingenious simulation games have been developed for modeling various aspects of life, but Eleusis is of special interest to mathematicians and scientists because it provides a model of induction, the process at the very heart of the scientific method. My first column on Eleusis was reprinted in *The 2nd Scientific American Book of Mathematical Puzzles & Diversions* (Simon and Schuster, 1961). Since then Eleusis has evolved into a game so much more exciting to play than the original version that I feel I owe it to readers to bring them up to date. I will begin, however, with some history.

Eleusis was invented in 1956 by Robert Abbott of New York, who at the time was an undergraduate at the University of Colorado. He had been studying that sudden insight into the solution of a problem that psychologists sometimes call the "Aha" reaction. Great turning points in science often hinge on these mysterious intuitive leaps. Eleusis turned out to be a fascinating simulation of this facet of science, even though Abbott did not invent it with this in mind. In 1963 Abbott's complete rules for the game appeared in his book, *Abbott's New Card Games* (hardcover, Stein & Day; paperback, Funk & Wagnalls).

Martin D. Kruskal, a distinguished mathematical physicist at Princeton University, became interested in the game and made several important improvements. In 1962 he published his rules in a monograph titled *Delphi: A Game of Inductive Reasoning.* Many college professors around the country used Eleusis and Delphi to explain scientific method to students and to model the Aha process. Artificial intelligence scientists wrote computer programs for the game. At the System Development Corporation in Santa Monica, research was done on Eleusis under the direction of J. Robert Newman. Litton Industries based a full-page advertisement on Eleusis. Descriptions of the game appeared in European books and periodicals. Abbott began receiving letters from all over the world with suggestions on how to make Eleusis a more playable game.

In 1973 Abbott discussed the game with John Jaworski, a young British mathematician who had been working on a computer version of Eleusis for teaching induction. Then Abbott embarked on a three-year program to reshape Eleusis, incorporating all the good suggestions he could. The new game is not only more exciting, its metaphorical level has been broadened as well. With the introduction of the roles of Prophet and False Prophet the game now simulates the search for any kind of truth. Here, then, based on a communication from Abbott, are the rules of New Eleusis as it is now played by aficionados.

At least four players are required. As many as eight can play, but beyond that the game becomes too long and chaotic.

Two standard decks, shuffled together are used. (Occasionally a round will continue long enough to require a third deck.) A full game consists of one or more rounds (hands of play) with a different player dealing each round. The dealer may be called by such titles as God, Nature, Tao, Brahma, the Oracle (as in Delphi) or just Dealer.

The dealer's first task is to make up a "secret rule." This is simply a rule that defines what cards can be legally played during a player's turn. In order to do well, players must figure out what the rule is. The faster a player discovers the rule, the higher his score will be.

One of the cleverest features of Eleusis is the scoring (described below), which makes it advantageous to the dealer to invent a rule that is neither too easy to guess nor too hard. Without this feature dealers would be tempted to formulate such complex rules that no one would guess them, and the game would become dull and frustrating.

An example of a rule that is too simple is: "Play a card of a color different from the color of the last card played." The alternation of colors would be immediately obvious. A better rule is: "Play so that primes and nonprimes alternate." For mathematicians, however, this might be too simple. For anyone else it might be too difficult. An example of a rule that is too complicated is: "Multiply the values of the last 3 cards played and divide by 4. If the remainder is 0, play a red card or a card with a value higher than 6. If the remainder is 1, play a black card or a picture card. If the remainder is 2, play an even card or a card with a value lower than 6. If the remainder is 3, play an odd card or a 10." No one will guess such a rule, and the dealer's score will be low.

Here are three examples of good rules for games with inexperienced players:

1. If the last legally played card was odd, play a black card. Otherwise play a red one.

2. If the last legally played card was black, play a card of equal or higher value. If the last card played was red, play a card of equal or lower value. (The values of the jack, queen, king and ace are respectively 11, 12, 13 and 1.)

3. The card played must be either of the same suit or the same value as the last card legally played.

The secret rules must deal only with the sequence of legally played cards. Of course, advanced players may use rules that refer to the entire pattern of legal and illegal cards on the table, but such rules are much harder to guess and are not allowed in standard play. Under no circumstances should the secret rule depend on circumstances external to the cards. Examples of such improper rules are those that depend on the sex of the last player, the time of day, whether God scratches his (or her) ear and so on.

The secret rule must be written down in unambiguous language, on a sheet of paper that is put aside for future confirmation. As Kruskal proposed, the dealer may give a truthful hint before the play begins. For example, he may say "Suits are irrelevant to the rule," or "The rule depends on the two previously played cards."

After the secret rule has been recorded, the dealer shuffles the double deck and deals 14 cards to each player and none to himself. He places a single card called the "starter" at the extreme left of the playing surface, as is indicated in Figure 105. To determine who plays first the dealer counts clockwise around the circle of players, starting with the player on his left and excluding himself. He counts until he reaches the number on the starter card. The player indicated at that number begins the play that then continues clockwise around the circle.

A play consists of placing one or more cards on the table. To play a single card the player takes a card from his hand and shows it to everyone. If according to the rule the card is playable, the dealer says "Right." The card is then placed to the right of the starter card, on the "main line" of correctly played cards extending horizontally to the right.

If the card fails to meet the rule, the dealer says "Wrong." In this case the card is placed directly below the last card played. Vertical columns of incorrect cards are called "sidelines." (Kruskal introduced both the layout and the terminology of the main line and sidelines.) Thus consecutive incorrect plays extend the same sideline downward. If a player displays a wrong card, the dealer gives him two more cards as a penalty, thereby increasing his hand.

If a player thinks he has discovered the secret rule, he may play a "string" of 2, 3 or 4 cards at once. To play a string he overlaps the cards slightly to preserve their order and shows them to everyone. If all the cards in the string conform to the rule, the dealer says "Right." Then all

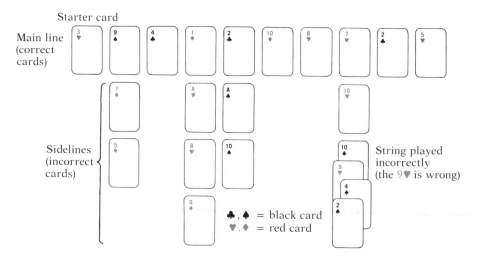

**Figure 105** *A typical round of Eleusis at an early stage*

the cards are placed on the main line with no overlapping, as if they were correctly played single cards.

If one or more cards in a string are wrong, the dealer declares the entire string wrong. He does not indicate which cards do not conform to the rule. The wrong cards are left overlapping to keep their identity as a string and the entire string goes below the last card played. The player is then dealt twice as many cards as there are in the string.

The layout shown in Figure 105 demonstrates all the rules of Eleusis mentioned so far. The dealer's secret rule for this layout is the first of the three given above.

Players improve their score by getting rid of as many cards as possible, and of course they can do this best if they guess the secret rule. At the start of a round there is little information to go on, and plays are necessarily random. As the round continues and more and more information is added to the layout, the rule becomes steadily easier to guess.

It may happen that a player thinks he knows the secret rule but finds he has no card that can be legally played. He then has the option of declaring "No play." In this case he shows his hand to everyone. If the dealer declares him right and his hand contains four cards or less, the cards are returned to the deck and the round ends. If he is right and has five or more cards, then his cards are put back into the deck, and he is dealt a fresh hand with four fewer cards than he previously held.

If the player is wrong in declaring no play, the dealer takes one of his correct cards and puts it on the main line. The player keeps the rest of his hand and, as a penalty, is dealt five more cards. A player who thinks he has no correct play but has not figured out the secret rule should realize that the odds are against his using the no play option successfully. He would do better to play a card at random.

When a player thinks he knows the secret rule, he has the opportunity to prove it and increase his score. He does so by declaring himself a Prophet. The Prophet immediately takes over the dealer's duties, calling plays right or wrong and dealing penalty cards when the others play. He can declare himself a Prophet only if all the following conditions prevail:

1. He has just played (correctly or incorrectly), and the next player has not played.
2. There is not already a Prophet.
3. At least two other players besides himself and the dealer are still in the round.
4. He has not been a Prophet before in this round.

When a player declares himself a Prophet, he puts a marker on the last card he played. A chess king or queen may be used. The Prophet keeps his hand but plays no more cards unless he is overthrown. The play continues to pass clockwise around the players' circle, skipping the Prophet.

Each time a player plays a card or string, the Prophet calls the play right or wrong. The dealer then either validates or invalidates the Prophet's statement by saying "Correct" or "Incorrect." If the Prophet is correct, the card or string is placed on the layout — on the main line if right or on a sideline if wrong — and the Prophet gives the player whatever penalty cards are required.

If the dealer says "Incorrect," the Prophet is instantly overthrown. He is declared a False Prophet. The dealer removes the False Prophet's marker and gives him five cards to add to his hand. He is not allowed to become a Prophet again during the same round, although any other player may do so. The religious symbolism is obvious, but as Abbott points out, there is also an amusing analogy here with science: "The Prophet is the scientist who publishes. The False Prophet is the scientist who publishes too early." It is the fun of becoming a Prophet and of overthrowing a False Prophet that is the most exciting feature of New Eleusis.

After a Prophet's downfall the dealer takes over his former duties. He completes the play that overthrew the Prophet, placing the card or string in its proper place on the layout. If the play is wrong, however, no penalty cards are given. The purpose of this exemption is to encourage players to make unusual plays — even deliberately wrong ones — in the hope of overthrowing the Prophet. In Karl Popper's language, it encourages scientists to think of ways of "falsifying" a colleague's doubtful theory.

If there is a Prophet and a player believes he has no card to play, things get a bit complicated. This seldom happens, and so you can skip this part of the rules now and refer to it only when the need arises. There are four possibilities once the player declares no play:

1. Prophet says, "Right"; dealer says, "Correct." The Prophet simply follows the procedure described earlier.

2. Prophet says, "Right"; dealer says, "Incorrect." The Prophet is immediately overthrown. The dealer takes over and handles everything as usual, except that the player is not given any penalty cards.

3.  Prophet says, "Wrong"; dealer says, "Incorrect." In other words, the player is right. The Prophet is overthrown, and the dealer handles the play as usual.

4.  Prophet says, "Wrong"; dealer says, "Correct." In this case the Prophet now must pick one correct card from the player's hand and put it on the main line. If he does this correctly, he deals the player the five penalty cards and the game goes on. It is possible, however, for the Prophet to make a mistake at this point and pick an incorrect card. If that happens, the Prophet is overthrown. The wrong card goes back into the player's hand and the dealer takes over with the usual procedure, except that the player is not given penalty cards.

After 30 cards have been played and there is no Prophet in the game, players are expelled from the round when they make a wrong play, that is, if they play a wrong card or make a wrong declaration of no play. An expelled player is given the usual penalty cards for his final play and then drops out of the round, retaining his hand for scoring.

If there is a Prophet, expulsions are delayed until at least 20 cards have been laid down after the Prophet's marker. Chess pawns are used as markers so that it is obvious when expulsion is possible. As long as there is no Prophet, a white pawn goes on every 10th card placed on the layout. If there is a Prophet, a black pawn goes on every 10th card laid down after the Prophet's marker. When a Prophet is overthrown, the black pawns and the Prophet's marker are removed.

A round can therefore go in and out of the phase when expulsions are possible. For example, if there are 35 cards on the layout and no Prophet, Smith is expelled when he plays incorrectly. Next Jones plays correctly and declares herself a Prophet. If Brown then plays incorrectly, she is not expelled because 20 cards have not yet been laid down after the Prophet's marker.

A round can end in two ways: (1) when a player runs out of cards or (2) when all players (excluding a Prophet, if there is one) have been expelled.

The scoring in Eleusis is as follows:

1.  The greatest number of cards held by anyone (including the Prophet) is called the "high count." Each player (including the Prophet) subtracts the number of cards in his hand from the high count. The difference is his score. If he has no cards, he gets a bonus of four points.

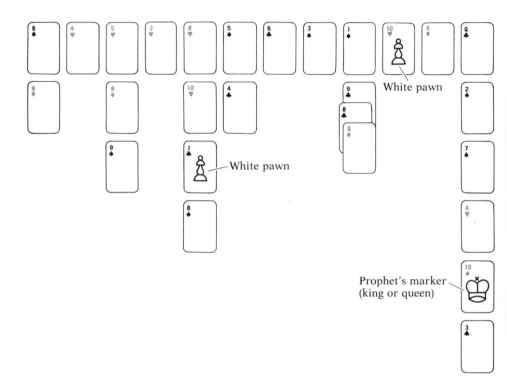

2. The Prophet, if there is one, also gets a bonus. It is the number of main-line cards that follow his marker plus twice the number of sideline cards that follow his marker, that is, a point for each correct card since he became a Prophet and two points for each wrong card.

3. The dealer's score equals the highest score of any player. There is one exception: If there is a Prophet, count the number of cards (right and wrong) that precede the Prophet's marker and double this number; if the result is smaller than the highest score, the dealer's score is that smaller number.

If there is time for another round, a new dealer is chosen. In principle the game ends after every player has been dealer, but this could take most of a day. To end the game before everyone has dealt, each player adds his scores for all the rounds played plus 10 more points if he has not been a dealer. This compensates for the fact that dealers tend to have higher-than-average scores.

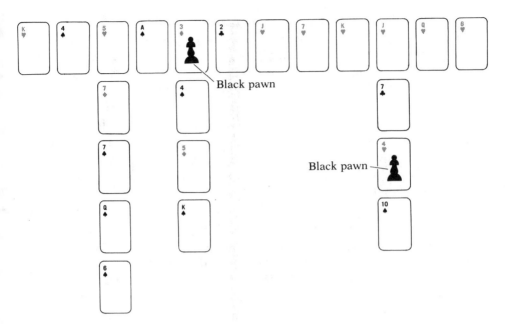

**Figure 106** *Layout at the end of a round of Eleusis includes a main line, several sidelines and various markers. White chess pawns are placed on every tenth card played in the round and black pawns on every tenth card played after a Prophet's marker*

The layout in Figure 106 shows the end of a round with five players. Smith was the dealer. The round ended when Jones got rid of her cards. Brown was the Prophet and ended with 9 cards. Robinson was expelled when he incorrectly played the 10 of spades; he had 14 cards. Adams had 17 cards at the end of the game.

The high count is 17. Therefore Adams' score is 17 minus 17, or 0. Robinson's score is 17 minus 14, or 3. Jones receives 17 minus 0, or 17, plus a 4-point bonus for having no cards, so that her score is 21. Brown gets 17 minus 9, or 8, plus the Prophet's bonus of 34 (12 main-line and 11 sideline cards following his marker), making a total score of 42. This is the highest score for the round. Twice the number of cards preceding the Prophet's marker is 50. Smith, the dealer, receives 42 because it is the smaller of the two numbers 42 and 50.

Readers are invited to look over this layout and see if they can guess the secret rule. The play has been standard, and so that the rule is confined strictly to the main-line sequence. I shall give the secret rule in the answer section.

Some miscellaneous advice from Abbott should help inexperienced Eleusis players. Since layouts tend to be large, the best way to play the game is on the floor. Of course a large table can be used as well as miniature cards on a smaller table. If necessary, the main line can be broken on the right and continued below on the left.

Remember that in Eleusis the dealer maximizes his score by choosing a rule that is neither too easy nor too difficult. Naturally this depends both on how shrewdly the dealer estimates the ability of the players and how accurately he evaluates the complexity of his rule. Both estimates require considerable experience. Beginning players tend to underestimate the complexity of their rules.

For example, the rule used in the first layout is simple. Compare it with: "Play a red card, then a black card, then an odd card, then an even card and repeat cyclically." This rule seems to be simpler, but in practice the shift from the red-black variable to the even-odd variable makes it difficult to discover. Abbott points out that in general restrictive rules that allow only about a fourth of the cards to be acceptable on any given play are easier to guess than less restrictive rules that allow half or more of the cards to be acceptable.

I shall not belabor the ways in which the game models a search for truth (scientific, mathematical or metaphysical) since I discussed them in my first column on the game. I shall add only the fantasy that God or Nature may be playing thousands, perhaps a countless number, of simultaneous Eleusis games with intelligences on planets in the universe, always maximizing his or her pleasure by a choice of rules that the lesser minds will find not too easy and not too hard to discover if given enough time. The supply of cards is infinite, and when a player is expelled, there are always others to take his place.

Prophets and False Prophets come and go, and who knows when one round will end and another begin? Searching for any kind of truth is an exhilarating game. It is worth remembering that there would be no game at all unless the rules were hidden.

## ANSWERS

The problem was to guess the secret rule that determined the final layout for a round in the card game Eleusis. The rule was: "If the last card is lower than the preceding legally played card, play a card higher than the last card, otherwise play a lower one. The first card played is correct unless it is equal to the starter card."

## ADDENDUM

Two unusual and excellent induction games have been invented since Eleusis and Delphi, both with strong analogies to scientific method. For Sid Sackson's board game Patterns, see Chapter 4 of my *Mathematical Circus* (Knopf, 1979). Pensari, using 32 special cards, was the brainchild of Robert Katz. His *Pensari Guide Book* (1986), which accompanies the cards, runs to 42 pages. Full-page advertisements for Pensari appeared in *Science News* in a number of 1987 issues.

A few years ago, reading *The Life and Letters of Thomas H. Huxley*, edited by his son Leonard (vol. 1, Appleton, 1901, page 262), I came across the following delightful paragraph. It is from a letter Huxley sent to Charles Kingsley in 1863.

> This universe is, I conceive, like to a great game being played out, and we poor mortals are allowed to take a hand. By great good fortune the wiser among us have made out some few of the rules of the game, as at present played. We call them "Laws of Nature," and honour them because we find that if we obey them we win something for our pains. The cards are our theories and hypotheses, the tricks our experimental verifications. But what sane man would endeavour to solve this problem: given the rules of a game and the winnings, to find whether the cards are made of pasteboard or goldleaf? Yet the problem of the metaphysicians is to my mind no saner.

## BIBLIOGRAPHY

*Delphi: A Game of Inductive Reasoning.* Martin D. Kruskal. Plasma Physics Laboratory, Princeton University, 1962. A monograph of 16 pages.

*The New Eleusis.* Robert Abbott. Privately published, 1977.

"Eleusis: The Game with the Secret Rule." Sid Sackson, in *Games*, May/June 1978, pp. 18–19.

"Simulating Scientific Inquiry with the Card Game Eleusis." H. Charles Romesburg, in *Science Education*, 3, 1979, pp. 599–608.

"The Methodology of Knowledge Layers for Inducing Descriptions of Sequentially Ordered Events." Thomas Dietterich. Department of Computer Science, University of Illinois at Urbana, master's thesis, 1980.

*New Rules for Classic Games.* R. Wayne Schmittberger. Wiley, 1992.

# CHAPTER 17

# Ramsey Theory

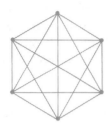

Prove that at a gathering of any six people, some
three of them are either mutual acquaintances
or complete strangers to each other.

—PROBLEM E 1321, *The American Mathematical
Monthly,* June–July, 1958

This chapter was originally written in 1977 to honor the
appearance of *The Journal of Graph Theory,* a periodical devoted to one
of the fastest growing branches of modern mathematics. Frank Harary,
the founding editor, is the author of the world's most widely used introduction to the subject. The current managing editor is Fan Chung of Bell
Communications Research.

Graph theory studies sets of points joined by lines. Two articles in the
first issue of the new journal dealt with Ramsey graph theory, a topic that
has a large overlap with recreational mathematics. Although a few
papers on Ramsey theory, by the Hungarian mathematician Paul Erdös
and others, appeared in the 1930's, it was not until the late 1950's that
work began in earnest on the search for what are now called Ramsey
numbers. One of the great stimulants to this search was the innocent-
seeming puzzle quoted above. It was making the mathematical-folklore
rounds as a graph problem at least as early as 1950, and at Harary's

suggestion it was included in the William Lowell Putnam Mathematical Competition of 1953.

It is easy to transform this puzzle into a graph problem. Six points represent the six people. Join every pair of points with a line, using a red pencil, say, to indicate two people who know each other and a blue pencil for two strangers. The problem now is to prove that no matter how the lines are colored, you cannot avoid producing either a red triangle (joining three mutual acquaintances) or a blue triangle (joining three strangers).

Ramsey theory, which deals with such problems, is named for an extraordinary University of Cambridge mathematician, Frank Plumpton Ramsey. Ramsey was only 26 when he died in 1930, a few days after an abdominal operation for jaundice. His father, A. S. Ramsey, was president of Magdalene College, Cambridge, and his younger brother Michael was archbishop of Canterbury from 1961 to 1974. Economists know him for his remarkable contributions to economic theory. Logicians know him for his simplification of Bertrand Russell's ramified theory of types (it is said that Ramsey Ramseyfied the ramified theory) and for his division of logic paradoxes into logical and semantical classes. Philosophers of science know him for his subjective interpretation of probability in terms of beliefs and for his invention of the "Ramsey sentence," a symbolic device that greatly clarifies the nature of the "theoretical language" of science.

In 1928 Ramsey read to the London Mathematical Society a now classic paper, "On a Problem of Formal Logic." (It is reprinted in *The Foundations of Mathematics*, a posthumous collection of Ramsey's essays edited by his friend R. B. Braithwaite.) In this paper Ramsey proved a deep result about sets that is now known as Ramsey's theorem. He proved it first for infinite sets, observing this to be easier than his next proof, for finite sets. Like so many theorems about sets, it turned out to have a large variety of unexpected applications to combinatorial problems. The theorem in its full generality is too complicated to explain here, but for our purposes it will be sufficient to see how it applies to graph-coloring theory.

When all pairs of $n$ points are joined by lines, the graph is called a complete graph on $n$ points and is symbolized by $K_n$. Since we are concerned only with topological properties, it does not matter how the points are placed or the lines are drawn. Figure 107 shows the usual ways of depicting complete graphs on two through six points. The lines identify every subset of $n$ that has exactly two members.

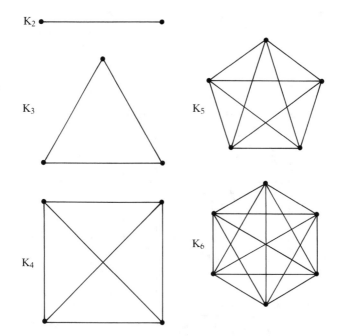

$K_2$

$K_3$

$K_4$

$K_5$

$K_6$

**Figure 107**   *Complete graphs on two through six points*

Suppose we arbitrarily color the lines of a $K_n$ graph red or blue. We might color the lines all red or all blue or any mixture in between. This is called a two-coloring of the graph. The coloring is of course a simple way to divide all the two-member subsets of $n$ into two mutually exclusive classes. Similarly, a three-coloring of the lines divides them into three classes. In general, an $r$-coloring divides the pairs of points into $r$ mutually exclusive classes.

A "subgraph" of a complete graph is any kind of graph contained in the complete graph in the sense that all the points and lines of the subgraph are in the larger graph. It is easy to see that any complete graph is a subgraph of any complete graph on more points. Many simple graphs have names. Figure 108 shows four families: paths, cycles, stars and wheels. Note that the wheel on four points is another way of drawing $K_4$. It is often called a tetrahedron because it is a planar projection of the tetrahedron's skeleton.

Consider now the following problem involving six pencils of different colors. To each color we assign any kind of graph we like. For example:

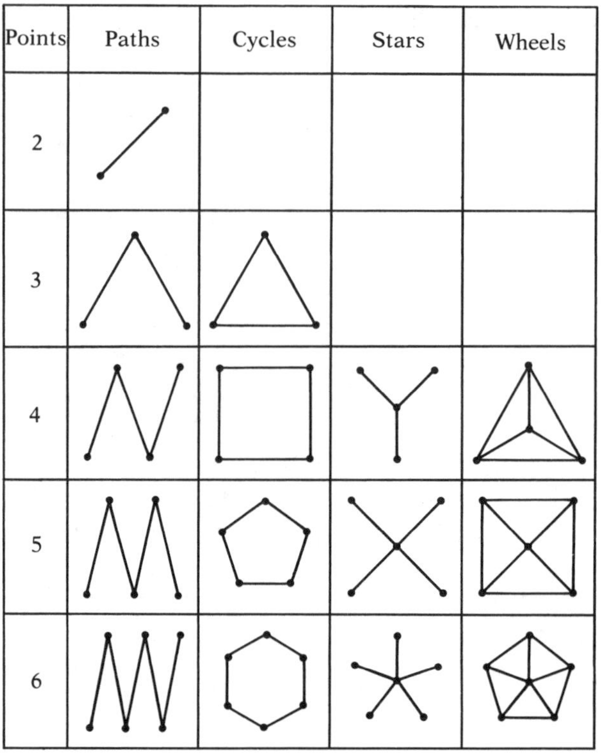

| Points | Paths | Cycles | Stars | Wheels |
|--------|-------|--------|-------|--------|
| 2 | / | | | |
| 3 | /\ | △ | | |
| 4 | N | □ | Y | △ |
| 5 | M | ⬠ | X | ⊠ |
| 6 | M | ⬡ | ✳ | ⬠ |

**Figure 108**   *Four important families of simple graphs*

1. Red: a pentagon (five-point cycle)
2. Orange: a tetrahedron
3. Yellow: a seven-point star
4. Green: a 13-point path
5. Blue: an eight-point wheel
6. Purple: a bow tie (two triangles sharing just one point)

We now ask a curious question. Are there complete graphs that, if their lines are arbitrarily six-colored, are certain to contain as a subgraph at least one of the six graphs listed above? In other words, no matter how we color one of these complete graphs with the six pencils are we certain to get either a red pentagon or an orange tetrahedron or a yellow seven-point star, and so on? Ramsey's theorem proves that beyond a certain size all complete graphs have this property. Let's call the smallest graph of this infinite set the Ramsey graph for the specified set of subgraphs. Its number of points is called the Ramsey number for that set of subgraphs.

Every Ramsey graph provides both a game and a puzzle. For our example, the game is as follows. Two players take turns picking up any one of the six pencils and coloring a line of the Ramsey graph. The first person to complete the coloring of one of the specified subgraphs is the loser. Since it is a Ramsey graph, the game cannot be a draw. Moreover, it is the smallest complete graph on which a draw is not possible.

The related puzzle involves a complete graph with one fewer point than the Ramsey graph. This obviously is the largest complete graph on which the game can be a draw. Such a graph is called the critical coloring for the specified set of subgraphs. The puzzle consists in finding a coloring for the critical graph in which none of the subgraphs appears.

I have no idea what the Ramsey number is for the six subgraphs given. Its complete graph would be so large (containing hundreds of points) that playing a game on it would be out of the question, and the associated puzzle is far too difficult to be within the range of a feasible computer search. Nevertheless, Ramsey games and puzzles with smaller complete graphs and with pencils of just two colors can be quite entertaining.

The best-known Ramsey game is called Sim after mathematician Gustavus Simmons who was the first to propose it. (The game was discussed in a column reprinted as Chapter 9 in my *Knotted Doughnuts and Other Mathematical Entertainments*.) Sim is played on the complete graph with six points ($K_6$), which models the problem about the party of six people. It is not hard to prove that 6 is the Ramsey number for the following two subgraphs:

1. Red: triangle ($K_3$)
2. Blue: triangle ($K_3$)

In "classical" Ramsey theory it is customary to use solitary numbers for complete graphs, and so we can express the above result with this compact notation: $R(3,3) = 6$. The $R$ stands for Ramsey number, the first 3 for a triangle of one color (say, red) and the second 3 for a triangle of another color (say, blue). In other words, the smallest complete graph that forces a "monochromatic" (all red or all blue) triangle when the graph is two-colored is 6. Thus if two players alternately color the $K_6$ red and blue, one player is certain to lose by completing a triangle of his color. The corresponding and easy puzzle is to two-color the critical graph, $K_5$, so that no monochromatic triangle appears.

It turns out that when $K_6$ is two-colored, at least two monochromatic triangles are forced. (If there are exactly two and they are of opposite color, they form a bow tie.) This raises an interesting question. If a complete graph on $n$ points is two-colored, how many monochromatic triangles are forced? A. W. Goodman was the first to answer this in a 1959 paper, "On Sets of Acquaintances and Strangers at Any Party." Goodman's formula is best broken into three parts: If $n$ has the form $2u$, the number of forced monochromatic triangles is $\frac{1}{3}u(u-1)(u-2)$. If $n$ is $4u+1$, the number is $\frac{2}{3}2u(u-1)(4u+1)$. If $n$ is $4u+3$, it is $\frac{1}{3}2u(u+1)(4u-1)$. Thus for complete graphs of 6 through 12 points the numbers of forced one-color triangles are 2, 4, 8, 12, 20, 28 and 40.

Random two-coloring will usually produce more monochromatic triangles than the number forced. When the coloring of a Ramsey graph contains exactly the forced number of triangles and no more, it is called extremal. Is there always an extremal coloring in which the forced triangles are all the same color? (Such colorings have been called blue-empty, meaning that the number of blue triangles is reduced to zero.) In 1961 Léopold Sauvé showed that the answer is no for all odd $n$, except for $n = 7$. This suggests a new class of puzzles. For example, draw the complete graph on seven points. Can you two-color it so that there are no blue triangles and no more than four red triangles? It is not easy.

Very little is known about "classical" Ramsey numbers. They are the number of points in the smallest complete graph that forces a given set of smaller complete graphs. There is no known practical procedure for finding classical Ramsey numbers. An algorithm is known: One simply explores all possible colorings of complete graphs, going up the ladder until the Ramsey graph is found. This task grows so exponentially in difficulty and at such a rapid rate, however, that it quickly becomes computationally infeasible. Even less is known about who wins—the first player or the second—if a Ramsey game is played rationally. Sim has been solved (it is a second-player win), but almost nothing is known about Ramsey games involving larger complete graphs.

So far we have considered only the kind of Ramsey game that Harary calls an avoidance game. As he has pointed out, at least three other kinds of game are possible. For example, in an "achievement" game (along the lines of Sim) the first player to complete a monochromatic triangle wins. In the other two games the play continues until all the lines are colored, and then either the player who has the most triangles of his color or the player who has the fewest wins. These last two games are the most difficult to analyze, and the achievement game is the easiest. In what follows "Ramsey game" denotes the avoidance game.

Apart from $R(3,3) = 6$, the basis of Sim, only the following seven other nontrivial classical Ramsey numbers are known for two-colorings:

1. $R(3,4) = 9$. If $K_9$ is two-colored, it forces a red triangle ($K_3$) or a blue tetrahedron ($K_4$). No one knows who wins if this is played as a Ramsey game.

2. $R(3,5) = 14$.

3. $R(4,4) = 18$. If $K_{18}$ is two-colored, a monochromatic tetrahedron ($K_4$) is forced. This is not a bad Ramsey game, although the difficulty of identifying tetrahedrons makes it hard to play. The graph and its coloring correspond to the fact that at a party of 18 people there is either a set of four acquaintances or four total strangers.

4. $R(3,6) = 18$. At the same party there is either a set of three acquaintances or six total strangers. In coloring terms, if a complete graph of 18 points is two-colored it forces either a red triangle or a blue complete graph of six points.

5. $R(3,7) = 23$.

6. $R(3,8) = 28$.

7. $R(3,9) = 36$.

8. $R(4,5) = 25$.

9. $R(6,7) = 298$.

As of March 1996, here are known bounds for eight other Ramsey numbers:

$R(3,10) = 40–43$.
$R(4,6) = 35–41$.
$R(4,7) = 49–61$.
$R(5,5) = 43–49$.
$R(5,6) = 58–87$.
$R(5,7) = 80–143$.
$R(6,6) = 102–165$.
$R(7,7) = 205–540$.

What is the smallest number of people that must include either a set of five acquaintances or a set of five strangers? This is equivalent to asking for the smallest complete graph that cannot be two-colored without producing a monochromatic complete graph with five points, which is the same as asking for the Ramsey number of R(5,5). So great is the jump in complexity from R(4,4) to R(5,5) that Stefan Burr, a leading expert on Ramsey theory who now teaches computer science at City College of the CUNY in New York City, thinks it possible that the number will never be known. Even R(4,5), he believes, is so difficult to analyze that it is possible its value also may never be found.

Only one other classical Ramsey number is known, and it is for three colors. R(3) = 3 is trivial because if you one-color a triangle, you are sure to get a one-color triangle. We have seen that R(3,3) equals 6. R(3,3,3) equals 17. This means that if $K_{17}$ is three-colored, it forces a monochromatic triangle. Actually it forces more than one, but the exact number is not known.

R(3,3,3) = 17 was first proved in 1955. The Ramsey game for this graph uses pencils of three different colors. Players alternately color a line, using any color they want to, until a player loses by completing a monochromatic triangle. Who wins if both players make their best possible moves? No one knows. The corresponding Ramsey puzzle is to three-color $K_{16}$, the critical graph, so that no monochromatic triangle appears. Plate 4, reproduced from *Graphs and Hypergraphs* by the French graph theorist Claude Berge, shows one of the two essentially different solutions. ("Different" is used here in a combinatorial sense deeper than mere exclusion of rotations and reflections.)

What about R(3,3,3,3), the minimum complete graph that forces a one-color triangle when it is four-colored? It is unknown, although an upper bound of 64 was proved by Jon Folkman, a brilliant combinatorialist who committed suicide in 1964 at the age of 31, following an operation for a massive brain tumor. The best lower bound, 51, was established by Fan Chung, a mathematician at Bell Communications Research who gave the proof in her Ph.D. thesis.

Classical Ramsey theory generalizes in many fascinating ways. We have already considered the most obvious way: the seeking of what are called generalized Ramsey numbers for *r*-colorings of complete graphs that force graphs other than complete ones. Václav Chvátal and Harary were the pioneers in this territory, and Burr has been mining it for the past 15 years. Consider the problem of finding Ramsey numbers for minimum complete graphs that force a monochromatic star of *n* points. Harary and Chvátal were the first to solve it for two-coloring. In 1973 Burr and J. A. Roberts solved it for any number of colors.

Another generalized Ramsey problem is to find Ramsey numbers for two-colorings of $K_n$ that force a specified number of monochromatic "disjoint" triangles. (Triangles are disjoint if they have no common point.) In 1975 Burr, Erdös and J. H. Spencer showed the number to be $5d$, where $d$ is the number of disjoint triangles and is greater than two. The problem is unsolved for more than two colors.

The general case of wheels is not even solved for two colors. The Ramsey number for the wheel of four points, the tetrahedron, is, as we have seen, 18. The wheel of five points (a wheel with a hub and four spokes) was shown to have a Ramsey number of 15 by Tim Moon, a Nigerian mathematician. The six-point wheel is unsolved, although its Ramsey number is known to have bounds of 17 to 20 inclusive, and there are rumors of an unpublished proof that the value is 17.

Figure 109, a valuable chart supplied by Burr and published for the first time in my 1977 column, lists the 113 graphs with no more than six lines and no isolated points, all of which have known generalized Ramsey numbers. Note that some of these graphs are not connected. In such cases the entire pattern, either all red or all blue, is forced by the complete graph with the Ramsey number indicated.

Every item on Burr's chart is the basis of a Ramsey game and puzzle, although it turns out that the puzzles — finding critical colorings for the critical graphs — are much easier than finding critical colorings for classical Ramsey numbers. Note that the chart gives six variations of Sim. A two-coloring of $K_6$ not only forces a monochromatic triangle but also forces a square, a four-point star (sometimes called a "claw"), a five-point path, a pair of disjoint paths of two and three points (both the same color), a square with a tail and the simple "tree" that is 15 on Burr's chart. The triangle with a tail (8), the five-point star (12), the Latin cross (27) and the fish (51) might be worth looking into as Ramsey games on $K_7$.

Ronald L. Graham of Bell Laboratories, one of the nation's top combinatorialists, has made many significant contributions to generalized Ramsey theory. It would be hard to find a creative mathematician who less resembles the motion-picture stereotype. In his youth Graham and two friends were professional trampoline performers who worked for a circus under the name of the Bouncing Baers. He is also one of the country's best jugglers and former president of the International Juggler's Association. The ceiling of his office is covered with a large net that he can lower and attach to his waist, so that when he is practicing with six or seven balls, any missed ball obligingly rolls back to him.

In 1968 Graham found an ingenious solution for a problem of the Ramsey type posed by Erdös and András Hajnal. What is the smallest graph of any kind, not containing $K_6$, that forces a monochromatic triangle when it is two-colored? Graham's unique solution is the eight-point graph shown in Figure 110. The proof is a straightforward *reductio*

| Graph Number | Ramsey Number | | Graph Number | Ramsey Number | | Graph Number | Ramsey Number |
|---|---|---|---|---|---|---|---|
| 1 | 2 | | 20 | 9 | | 39 | 9 |
| 2 | 3 | | 21 | 8 | | 40 | 9 |
| 3 | 6 | | 22 | 7 | | 41 | 11 |
| 4 | 5 | | 23 | 8 | | 42 | 10 |
| 5 | 5 | | 24 | 7 | | 43 | 11 |
| 6 | 6 | | 25 | 8 | | 44 | 10 |
| 7 | 6 | | 26 | 7 | | 45 | 12 |
| 8 | 7 | | 27 | 7 | | 46 | 14 |
| 9 | 10 | | 28 | 8 | | 47 | 9 |

| Graph Number | Ramsey Number | | Graph Number | Ramsey Number | | Graph Number | Ramsey Number |
|---|---|---|---|---|---|---|---|
| 58 | 11 | | 77 | 9 | | 96 | 10 |
| 59 | 11 | | 78 | 9 | | 97 | 11 |
| 60 | 11 | | 79 | 9 | | 98 | 10 |
| 61 | 11 | | 80 | 9 | | 99 | 11 |
| 62 | 8 | | 81 | 10 | | 100 | 11 |
| 63 | 10 | | 82 | 9 | | 101 | 11 |
| 64 | 11 | | 83 | 10 | | 102 | 13 |
| 65 | 10 | | 84 | 9 | | 103 | 12 |
| 66 | 11 | | 85 | 10 | | 104 | 12 |

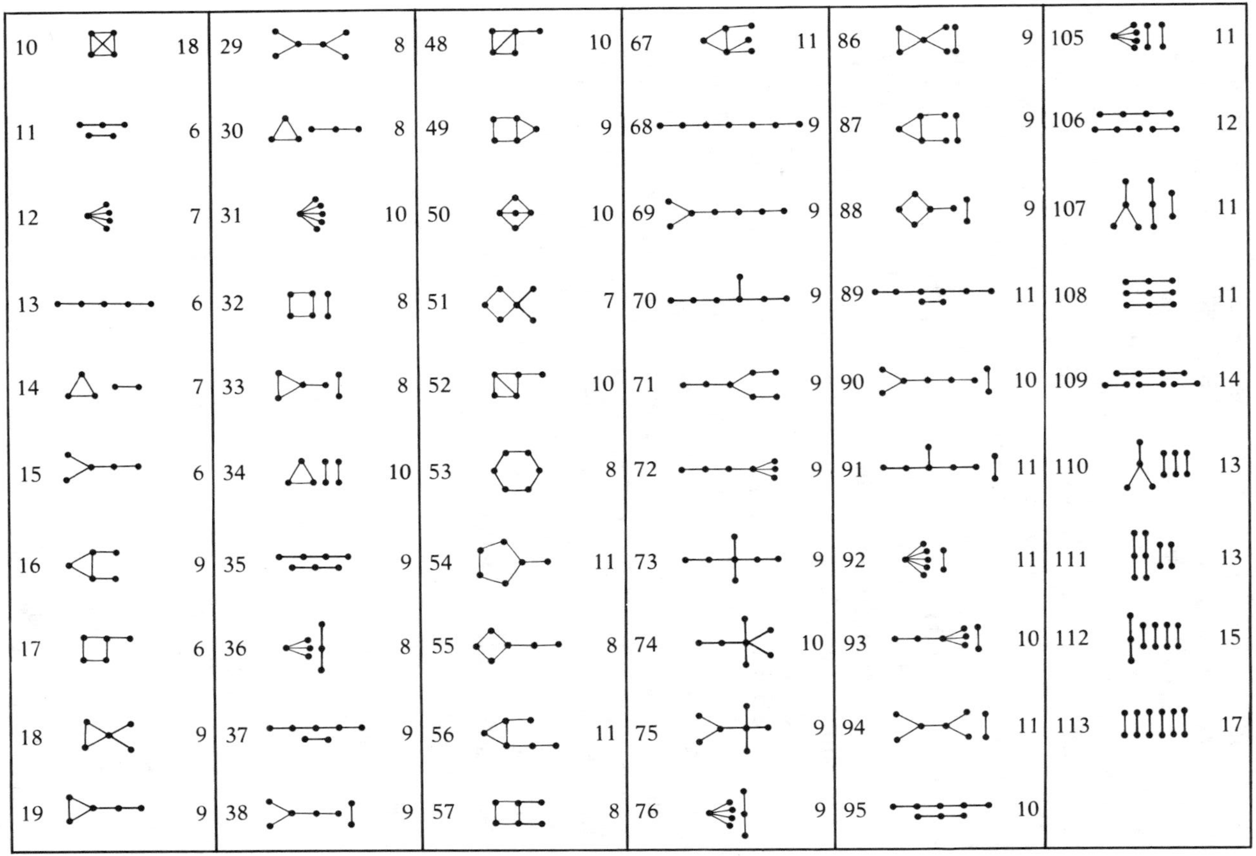

*Figure 109   Simple graphs for which the generalized Ramsey number is known*

*ad absurdum.* It begins with the assumption that a two-coloring that avoids monochromatic triangles is possible and then shows that this forces such a triangle. At least two lines from the top point must be, say, gray, and the graph's symmetry allows us to make the two outside lines gray with no loss of generality. The endpoints of these two lines must then be joined by a colored line (shown dotted) to prevent the formation of a gray triangle. Readers may enjoy trying to complete the argument.

What about similar problems when the excluded subgraph is a complete graph other than $K_6$? The question is meaningless for $K_3$ because $K_3$ is itself a triangle. $K_5$ is unsolved. The best-known solution is a 16-point graph discovered by two Bulgarian mathematicians. $K_4$ is even further from being solved. Folkman, in a paper published posthumously, proved that such a Ramsey graph exists, but his construction used more than $2 \uparrow \uparrow \uparrow 2^{901}$ points. This is such a monstrous number that there is no way to express it without using a special arrow notation. The notation is introduced by Donald E. Knuth in his article "Mathematics and Computer Science: Coping with Finiteness" in *Science* (December 17, 1976). The number has since been lowered to a more respectable size.

Imagine the universe tightly packed with spheres the size of electrons. The total number of such spheres is inconceivably smaller than the number occurring in Folkman's graph. Erdös has a standing offer of $100 to anyone finding a graph for this problem that has fewer than a million points.

Folkman's graph dramatically illustrates how enormously difficult a

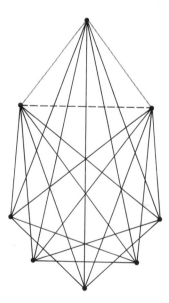

**Figure 110**
*Graham's solution to a problem by Erdös*

Ramsey problem can be even when the problem's statement mentions no graph with more than four points. But as Al Jolson liked to say, you ain't heard nothin' yet. Graham has found an even more mind-boggling example.

Consider a cube with lines joining every pair of corners. The result is a complete graph on eight points, except now we have added a Euclidean geometric structure. Imagine the lines of this spatial $K_8$ arbitrarily colored red and blue. Can it be done in such a way that no monochromatic $K_4$ results that *lies on a plane*? The answer is yes, and it is not hard to do.

Let us generalize to $n$-dimensional cubes. A hypercube has $2^n$ corners. On the four-dimensional hypercube, it also is possible to two-color the lines of the complete graph of $2^4$, or 16, points so that no one-color complete planar graph of four points results. The same can be done with the $2^5$ hypercube of 32 points. This suggests the following Euclidean Ramsey problem: What is the smallest dimension of a hypercube such that if the lines joining all pairs of corners are two-colored, a planar $K_4$ of one color will be forced? Ramsey's theorem guarantees that the question has an answer only if the forced $K_4$ is not confined to a plane.

The existence of an answer when the forced monochromatic $K_4$ is planar was first proved by Graham and Bruce L. Rothschild in a far-reaching generalization of Ramsey's theorem that they found in 1970. Finding the actual number, however, is something else. In an unpublished proof Graham has established an upper bound, but it is a bound so vast that it holds the record for the largest number ever used in a serious mathematical proof.

To convey at least a vague notion of the size of Graham's number we must first attempt to explain Knuth's arrow notation. The number written $3 \uparrow 3$ is $3 \times 3 \times 3 = 3^3 = 27$. The number $3 \uparrow \uparrow 3$ denotes the expression $3 \uparrow (3 \uparrow 3)$. Since $3 \uparrow 3$ equals 27, we can write $3 \uparrow \uparrow 3$ as $3 \uparrow 27$ or $3^{27}$. As a slanting tower of exponents it is

$$3^{3^3}$$

The tower is only three levels high, but written as an ordinary number it is 7,625,597,484,987. This is a big leap from 27, but it is still such a small number that we can actually print it.

When the huge number $3 \uparrow \uparrow \uparrow 3 = 3 \uparrow \uparrow (3 \uparrow \uparrow 3) = 3 \uparrow \uparrow 3^{27}$ is written as a tower of 3's, it reaches a height of 7,625,597,484,987 levels. Both the tower and the number it represents are now too big to be printed without special notation.

Consider $3 \uparrow \uparrow \uparrow \uparrow 3 = 3 \uparrow \uparrow \uparrow (3 \uparrow \uparrow \uparrow 3)$. Inside the parentheses is the gigantic number obtained by the preceding calculation. It is no longer possible to indicate in any simple way the height of the tower of 3's that expresses $3 \uparrow \uparrow \uparrow \uparrow 3$. The height is another universe away from $3 \uparrow \uparrow \uparrow \uparrow 3$. If we break $3 \uparrow \uparrow \uparrow \uparrow 3$ down to a series of the double-arrow operations, it is $3 \uparrow \uparrow (3 \uparrow \uparrow (3 \uparrow \uparrow \ldots \uparrow \uparrow (3 \uparrow \uparrow 3) \ldots ))$, where the number of steps to be iterated is $3 \uparrow \uparrow \uparrow 3$. As Knuth says, the dots "suppress a lot of detail." $3 \uparrow \uparrow \uparrow \uparrow 3$ is unimaginably larger than $3 \uparrow \uparrow \uparrow 3$, but it is still small as finite numbers go, since most finite numbers are very much larger.

We are now in a position to indicate Graham's number. It is represented in Figure 111. At the top is $3 \uparrow \uparrow \uparrow \uparrow 3$. This gives the number of arrows in the number just below it. That number in turn gives the number of arrows below it. This continues for $2^6$, or 64, layers. It is the bottom number that Graham has proved to be an upper bound for the hypercube problem.

Now hold on to your hat. Ramsey-theory experts believe the actual Ramsey number for this problem is probably 6. As Stanislaw M. Ulam said many times in his lectures, "The infinite we shall do right away. The finite may take a little longer."

## ANSWERS

Figure 112 (note the color symmetry) shows how to two-color the complete graph on seven points so that it is blue-empty (gray lines) and contains four (the minimum) red triangles (black lines). If you enjoyed

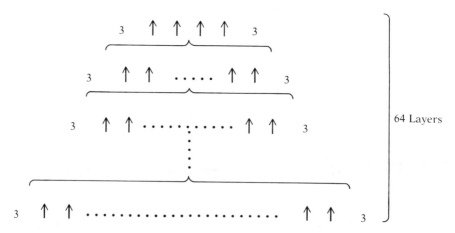

**Figure 111**  *Graham's upper bound for the solution to a Euclidean Ramsey problem*

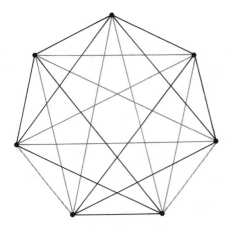

***Figure 112***
*Solution to a Ramsey graph puzzle*

working on this problem, you might like to tackle $K_8$ by two-coloring the complete graph on eight points so that it is blue-empty and has eight (the minimum) red triangles.

Garry Lorden, in his 1962 paper, "Blue-Empty Chromatic Graphs," showed that problems of this type are uninteresting. If the number of points $n$ is even, Goodman had earlier shown that the graph can be made blue-empty extremal simply by coloring red two complete subgraphs of $n/2$ points. If $n$ is odd, $K_7$ is (as I said earlier) the only complete graph that can be made blue-empty extremal by two-coloring. When two-coloring complete graphs where $n$ is odd and not 7, how do you create a blue-empty graph that is not extremal but has the smallest number of red triangles? Lorden showed that this is easily done by partitioning the graph into two complete red subgraphs, one of $(n + 1)/2$ points, the other of $(n - 1)/2$ points. Some extra red lines may be required, but their addition is trivial.

## ADDENDUM

Since this chapter first appeared in *Scientific American* in 1977, such rapid progress has been made in Ramsey theory that a bibliography of papers would run to more than a thousand titles and any effort to summarize here even the main results would be impossible. Fortunately, two excellent books are now available as introductions to the general field (of which Ramsey graph theory is only a part): Ronald Graham's *Rudiments of Ramsey Theory* and *Ramsey Theory*, a book Graham coauthored with two colleagues.

Ramsey theory now includes, among many other things, problems involving the partitioning of lines in any graph or the partitioning of points in any space. Euclidean Ramsey theory, pioneered in the 1970's by Graham, Erdös and others, concerns the $k$-coloring of all points in a given Euclidean space and determining what patterns are forced. For example, no matter how the points on the plane are two-colored, the coloring will force the vertices of a monochromatic triangle of any specified size and shape except the equilateral triangle. (The plane can be colored with stripes of two alternating colors of such widths that no equilateral triangle, say of side 1, can have all its corners the same color.)

If all the points in Euclidean three-dimensional space are two-colored, will it force an equilateral triangle of any size? Yes. Consider the four points of any regular tetrahedron. No matter how the points are two-colored, at least three must be the same color, and of course those three will form a monochromatic equilateral triangle. For other theorems of this type, much harder to prove, see the 1973 paper by Erdös, Graham and others.

During the past decade Frank Harary and his associates have been analyzing Ramsey games. Only a small fraction of their results have been published, but Harary is planning a large book on what he calls achievement and avoidance games of the Ramsey type.

In Stefan Burr's chart (Figure 109) showing all graphs with six or fewer lines, you will see that in the interval of integers 2 through 18, only 4 and 16 are missing as generalized Ramsey numbers. There are three graphs of seven lines each that have generalized Ramsey numbers of 16, but there is no graph that has a generalized Ramsey number of 4. Is every positive integer except 4 (ignoring 1, which is meaningless) a generalized Ramsey number? In 1970 Harary showed that the answer is yes. The impossibility of 4 is easily seen by inspecting Burr's chart. The tetrahedron (complete graph for four points) cannot be the smallest graph that forces a subgraph because every possible subgraph of a tetrahedron has a Ramsey number higher or lower than 4.

What is the smallest complete graph which, if two-colored, will force a monochromatic complete graph for five points? In other words, what is the generalized Ramsey number $R(5,5)$? In 1975 Harary offered $100 for the first solution, but the prize remains unclaimed.

"Ramsey theory is only in its infancy," writes Harary in his tribute to Ramsey in the special 1983 issue of *The Journal of Graph Theory* cited in the bibliography and adds: "There is no way that Frank Ramsey could have foreseen the theory his work has inspired." D. H. Mellor, in another tribute to Ramsey in the same issue, has this memorable sentence: "Ramsey's enduring fame in mathematics . . . rests on a theorem he didn't need, proved in the course of trying to do something we now know can't be done!"

## BIBLIOGRAPHY

"On Sets of Acquaintances and Strangers at Any Party." A. W. Goodman, in *The American Mathematical Monthly*, 66, 1959, pp. 778–783.

"Blue-Empty Chromatic Graphs." Garry Lorden, in *The American Mathematical Monthly*, 69, 1962, pp. 114–120.

"On Edgewise 2-Colored Graphs with Monochromatic Triangles and Containing No Complete Hexagon." Ronald Graham, in *Journal of Combinatorial Theory*, 4, 1968, p. 300.

*Graph Theory*. Frank Harary. Addison-Wesley, 1969.

"Euclidean Ramsey Theorems I." P. Erdös, et al., in *Journal of Combinatorial Theory*, Series A, 14, 1973, pp. 341–363.

"Generalized Ramsey Theory for Graphs—A Survey." Stefan Burr, in *Graphs and Combinatorics*. R. Bari and F. Harary, eds. Springer-Verlag, 1974, pp. 58–75.

*Ramsey Theory*. Ronald Graham, Bruce Rothschild and Joel Spencer. Wiley, 1980.

*Rudiments of Ramsey Theory*. Ronald Graham. *The American Mathematical Society*, 1981.

*The Journal of Graph Theory*, 7, Spring 1983. This issue, dedicated to Frank Ramsey, contains 19 papers about Ramsey and Ramsey theory.

"Ramsey Theory." Ronald Graham and Joel Spencer, in *Scientific American*, July 1990, pp. 112–117.

*Mathematics of Ramsey Theory*. J. Nesetril and V. Rodl (eds.). Springer, 1990.

"Party Numbers." Ivars Peterson, in *Science News,*, Vol. 144, July 17, 1993, pp. 46–47.

"People Who Know People." Michael Albertson, in *Mathematics Magazine*, Vol. 67, October 1994, pp. 278–81.

"A Visit to Asymptopia." Barry Cipra, in *Science*, Vol. 267, February 17, 1995, pp. 264–65.

CHAPTER **18**

# From Burrs
# to Berrocal

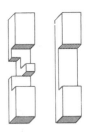

We then fell into a disquisition whether there is
any beauty independent of utility. The General
maintained there was not. Dr. Johnson
maintained that there was; and he instanced a
coffee-cup which he held in his hand, the
painting of which was of no real use, as the
cup would hold the coffee equally well if plain;
yet the painting was beautiful.

—JAMES BOSWELL, *Life of Samuel Johnson*

The purpose of this chapter is to introduce readers to
Miguel Berrocal, the foremost living Spanish sculptor. Berrocal is already a cult figure in Europe with a steadily growing following, but he is
surprisingly little known in the U.S. Before detailing Berrocal's remarkable achievement, however, and explaining why his sculpture should be
discussed in a book on recreational mathematics, we must consider one
of the oldest categories of mechanical puzzles.

Is there any reader who has not at some time held one of those puzzles made of wood pieces so cleverly interlocked that they are quite difficult to separate? These objects are usually called Chinese puzzles, and once they have been taken apart, putting them back together again can be even more challenging. There is usually a single piece, called the key, that must be removed first in order to disassemble the puzzle. In many such puzzles the pieces must be replaced in a certain sequence in which the key is always inserted last to lock the other pieces firmly in place. Hundreds of differently structured puzzles of this type have been sold around the world for centuries, most of them invented anonymously (even though hundreds of patents for unmarketed variations on the puzzles have been issued).

Almost nothing is known of the early history of these puzzles, but "Chinese puzzles" is probably a misnomer. East Asian countries were certainly producing puzzles of this kind as early as the eighteenth century, but so were European countries. Moreover, other kinds of mechanical puzzles of unquestionably Western origin were also called Chinese puzzles. As Joseph Needham writes in Volume III of *Science and Civilization in China:* "Perhaps Europeans were inclined to ascribe to puzzles the name of what was, to them, a puzzling civilization."

Chinese puzzles can have as few as three pieces, but the simplest nontrivial model is the popular six-piece puzzle shown in Figure 113.

*Figure 113* A six-piece "burr" puzzle

The six pieces are shown below the assembled puzzle as they were depicted by "Professor Hoffmann" (the pseudonym of Angelo Lewis) in his 1893 book on mechanical puzzles, *Puzzles Old and New*. The unnotched bar at the far left is the key. A different set of six pieces appeared earlier in *The Magician's Own Book*, an anonymous work first published in the U.S. in 1857. Anthony S. Filipiak, in *100 Puzzles: How to Make and Solve Them* (A. S. Barnes and Co., 1942), calls such an assembly a "six-piece burr puzzle," presumably because it looks like a seed burr. The name "burr" is now commonly applied to all puzzles of this kind.

Manufactured versions of the six-piece burr are seldom alike. This fact suggested to Filipiak a surprisingly difficult problem in geometric combinatorics. Imagine that the middle part of the puzzle's unnotched key piece is divided into 12 unit cubes as shown in Figure 114. The sides of each cube are equal to half the depth of the key. There are four more cubes under the cubes numbered 2, 3, 6 and 7, but only two of them, 11 and 12, can be seen in the illustration. Each piece of any six-piece burr can now be described by stating which of the 12 unit cubes have been removed.

Since each unit cube is either present or absent in a piece of the puzzle, there are $2^{12}$, or 4096, possible patterns in all. Those that divide the bar into separate pieces must of course be eliminated. Among the remaining pieces, simple rotations and end-for-end turns may account for as many as eight repetitions of any single pattern. Therefore for each pattern seven duplicates can be eliminated. (Mirror reflections of asymmetric configurations are not excluded.) Finally, there are some patterns that for mechanical reasons will not combine with any other pattern to

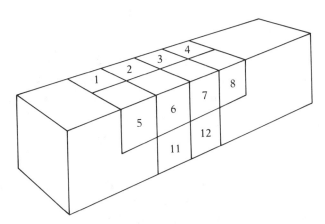

**Figure 114** *Pattern for the pieces of a burr puzzle*

form an interlocking structure. These patterns can also be eliminated. After the above exclusions have been made, how many configurations remain? In other words, how many of the 4096 patterns can be utilized in a six-piece burr puzzle? Filipiak believed there were 432.

In recent years three mathematicians have employed computer programs to attack the problem. They are William H. Cutler of Wartburg College in Waverly, Iowa, Robert H. Mackay of London and C. Arthur Cross of Cheshire in England. Cutler and Cross now agree that the number of usable pieces is 369. Of these, 112 can be utilized with duplicates and 2 can be utilized with triplicates, making a set of 485 pieces in all. Twenty-five of them are called "notchable" in the sense that they can be cut with a table saw and can be used to make puzzles without interior holes. Excluding interlockings that are impossible to put together or take apart (without going through four-dimensional space) and omitting structures with internal holes because they hold together poorly, how many distinct burr puzzles can be made with 6 pieces selected from the set of 485? Computer programs by Cutler and others have found the number to be 119,979.

I know of only two places where more elaborate kinds of burr puzzles can be obtained. In England, Pentangle (Over Wallop, Hants, England S020 8JA) will send a catalogue of their mechanical-puzzle line, which includes several handsome wood puzzles of the burr type. The one called "Grandpa Chuck" has 96 pieces. In the U.S. if readers send a stamped and self-addressed business envelope to Stewart T. Coffin (Old Sudbury Road, R.F.D. 1, Lincoln, MA 01773), he will send back a brochure for his unusual line of original burr puzzles. His prices are high because all the puzzles are handcrafted from hardwoods. For several years Coffin issued an occasional newsletter, *Puzzle Craft*, that dealt with the history and construction of mechanical puzzles.

From time to time Chinese puzzles have been manufactured to resemble familiar objects: a car, a pistol, a battleship, an airplane, a pagoda, a barrel, an egg and various animals. Most interlocking wood puzzles are not designed to represent anything, yet because of their symmetries they are always pleasing to look at. It is the representational models that are usually low in aesthetic appeal. This brings me to Berrocal. As far as I know, he is the first person to combine the interlocking Chinese puzzle with high art.

Berrocal was born in 1933 into a middle-class Spanish family in Málaga. He studied mathematics and architecture at the University of Madrid and then art in Paris and Rome before settling in Negrar, a suburb of Verona. He lives there today in a palatial villa with his second wife, Princess Cristina. (She is the granddaughter of the last king of Portugal.) Berrocal presides over a 200-employee foundry in Negrar that casts not only his own work but also much work of other European sculptors. "I am the boss of the sculptors' Mafia," he once said. Like Pablo Picasso and Salvador Dali, the most notable Spanish painters of the twentieth century, Berrocal is a virtuoso who combines a prodigious output with skillful public relations and a dazzling, immodest personality.

It is impossible to appreciate the unique combination of values in Berrocal's work—visual beauty, tactile pleasure, humor and the intellectual stimulation of a three-dimensional combinatorial puzzle—until one has taken a Berrocal apart and put it back together several times. For example, consider the two Berrocals shown stacked together in Figure 115. The head, called *Portrait de Michèle*, consists of 17 separate, curiously shaped elements, each one designed to be an individual piece of abstract sculpture and also pleasing to the fingers. The body, called *La Totoche* (The Plump Lady), is one of several bodies onto which the head can be fitted. It can be broken down into 12 pieces.

Berrocal introduced the term "multiples" for the mass-produced copies that he casts of each of his works. His edition of *Portrait de Michèle* is typical: 6 multiples in solid gold, 500 in sterling silver and 9500 in nickel-plated bronze. Each copy is numbered and signed and flawlessly crafted with engineering precision. In every Berrocal the pieces must be separated and reassembled in a certain order. To disassemble *Portrait de Michèle* one must first remove an element in the neck. Chinese burr puzzles tend to fall apart as soon as a few pieces are removed, but a Berrocal holds firmly together until the last two pieces are separated. In many cases the assembled model is completely solid— that is, there are no interior holes—and until piece $n$ is removed, it is impossible to remove piece $n + 1$. The interlockings are so ingenious that sometimes a piece cannot be taken out until the positions of other pieces have been slightly altered. Every multiple comes with a hardcover instruction book illustrated by Berrocal. Each stage of assembly is depicted on a separate page with the piece to be moved shown in color. At the end, an isometric drawing displays the outline of all the pieces (in transparent form) after they are in place. Even when the instruction book is consulted, however, it may take several days to master the technique of taking apart and reassembling a Berrocal.

The number of pieces in a Berrocal varies from 3 to almost 100. Exquisite finger rings and bracelets, all wearable, are elements in many of the sculptures. For example, the pupil of Michèle's eye is the aquama-

**Figure 115** *Portrait de Michèle* (head) *attached to La Totoche* (body)

rine stone of a ring. *Mini David*, the torso shown in Figure 116, left, is another popular Berrocal. One of its 22 pieces is the ring shown in Figure 117, top. The genitals of the torso hang below the gem of the ring. The entire edition of this work has been sold. It consists of 6 solid gold,

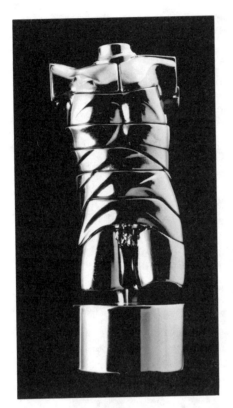 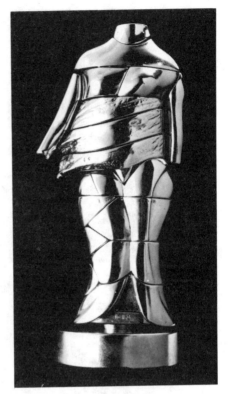

**Figure 116**  Mini David (left) and Mini Cariatide (right)

500 gold-plated bronze and 9500 nickel-plated bronze multiples. In the gold-plated set the gem of the ring is green jade; in the nickel-plated set it is sapphire. Berrocal often produces "micro" versions of his works designed as pendants. There is a *Micro David* that includes a ring with a mesh band and a blue heart of lapis lazuli. The internal structure of a micro work is always completely different from that of its mini counterpart.

*Mini Maria*, shown in Figure 118, can be disassembled only when a ball bearing on one leg is pressed. This sculpture is made of 23 pieces. One of them is a ring with a moonstone that forms one of Maria's breasts. Inside the figure there is a male sex organ, appearing at the left in Figure 119, that comes apart in five pieces, two of which are steel balls. The corresponding element in the pendant *Micro Maria* has a tiny aquamarine on the tip of the organ. The gem is attached to a ring with a mesh band.

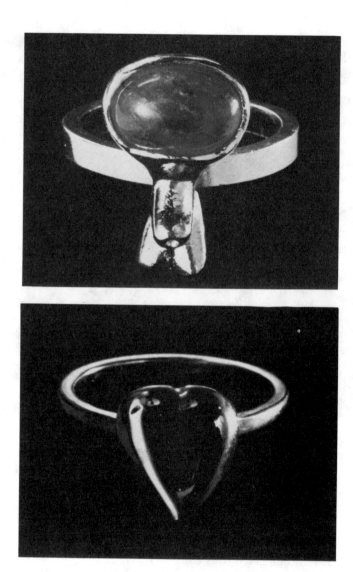

**Figure 117**   *Rings from Mini David* (top) *and Mini Cariatide* (bottom)

Another reclining figure, *Mini Zoriade*, opens when one shoe is rotated. Zoriade's breasts are the moonstones of a ring. *Mini Cariatide* appears at the right in Figure 116. The mons veneris of the figure is on the gold ring appearing at the bottom in Figure 117.

*Goliath*, shown in Figure 120, is Berrocal's most complicated work. All 80 of its elements appear in Figure 121. Figure 122, a partially disassembled *Richelieu*, shows how a Berrocal is a striking work of

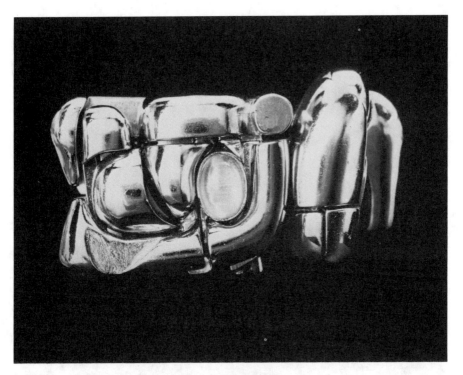

**Figure 118**  Berrocal's Mini Maria

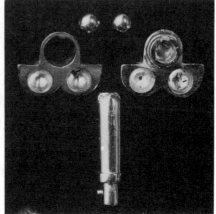

**Figure 119**  An interior element of Mini Maria (left) and its five pieces (right)

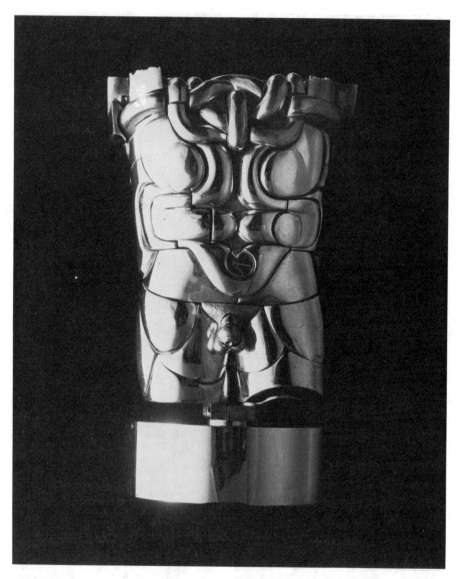

**Figure 120**  *Miguel Berrocal's Goliath*

abstract sculpture at any stage of its assembly. When the torso of *Goliath* is completely assembled, its fig leaf can be rotated to expose the genitalia. Actually there are two pieces representing genitalia, one circumcised and one not, and the torso can be assembled with either one exposed.

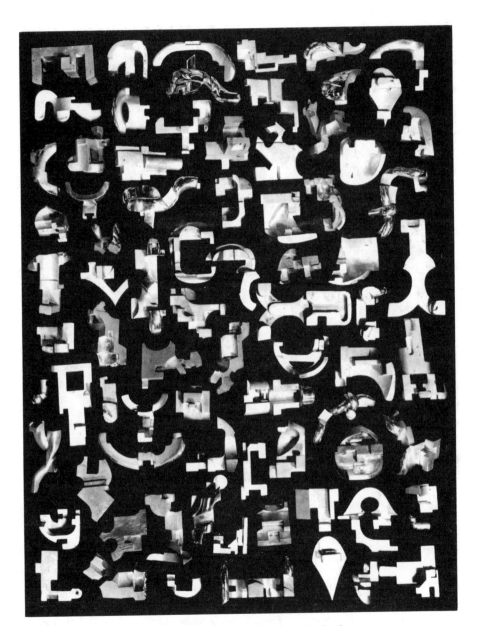

**Figure 121** *The 80 pieces of Goliath*

Berrocal's coffin for Romeo and Juliet appears in Figure 123, together with an earlier 16-piece work that depicts the ill-fated lovers intertwined. Inside the lovers is a surprise more suitably described in a sex manual. The coffin is even more surprising. It disassembles into 84

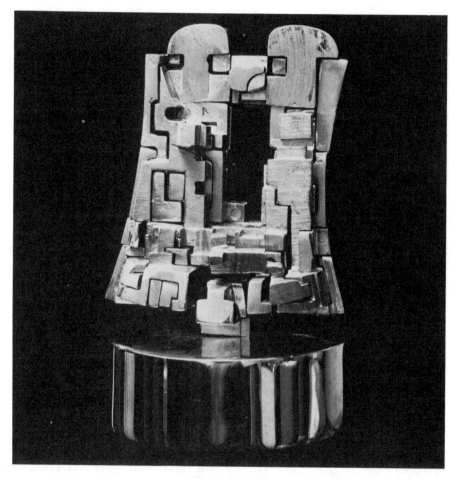

**Figure 122** *Richelieu partially disassembled*

pieces that can be refitted to make complete place settings for two: 23 pieces of silverware, 4 wine goblets, 4 candlesticks, 4 ashtrays, a man's dinner ring, a woman's dinner ring and a chafing dish that is larger than the coffin.

*Columbia Jet,* shown in Figure 124, was commissioned by Iberia Airlines as a gift for its executives. The bird is Picasso's dove of peace. Its body is a water pitcher that pours from the dove's beak when a handle is raised. The ornate pedestal is a drinking glass. Unfold the dove's wings and landing gears descend from the wing tips. Another useful Berrocal is *Paloma Box,* a sculpture that is about a foot high and wide and opens into a jewel case with 16 felt-lined drawers. A circular mirror can be raised and then opened to reveal a sculpture of the head of Picasso's daughter Paloma. The pieces making up the head include two bracelets and two belts. Berrocal's *Il Cavallo*, shown in Figure 125, has a different kind of flexibility. The 14-piece horse has legs articulated so that it can assume different positions.

I have described only a small portion of Berrocal's work. His largest sculpture is his *Homage to Picasso*, which is 18 feet long and weighs 18 tons. It is now on permanent display in the Picasso Gardens in Málaga. A miniature bronze version, *Siextasis*, consists of 20 pieces locked together by 20 small magnets. Another large sculpture, Berrocal's tribute

**Figure 123**  *A sculpture of Romeo and Juliet on top of another sculpture depicting their coffin*

**Figure 124** *Berrocal's Columbia Jet*

to his good friend Dali, is in Madrid. A miniature bronze version, called *Dalirium Tremens*, was made later.

There are no screws or bolts in any Berrocals except the pendants. The pendants hang by a key piece that screws in to prevent the pendant from accidentally dropping off its chain. All gold and sterling-silver mini multiples have key pieces that lock into place with an ordinary key so that the sculpture cannot be disassembled without the owner's cooperation.

Let me anticipate an objection. What, you may ask, does the fun of taking sculpture apart and putting it together again like a Chinese puzzle have to do with art? In a sense the answer is "Nothing," but that is not the whole story. The visual beauty of art has always been combined in countless ways with other values: the sexual emotions aroused by nudes; the sentiments evoked by landscapes, seascapes and family portraits; the rhetorical function of political and religious art; the didactic value of textbook illustration; the humor of comic art; the physical comfort of chairs, beds and sofas that are designed to be beautiful; the utility of tables, vases, bottles, cups, dishes, silverware, cars, houses, quilts, ships, watches, tools and so on. Berrocal's unique achievement is the combining of visual and tactile pleasures with the intellectual play of a mechanical puzzle. If you do not enjoy that particular combination, then a Berrocal is not for you.

In the Quadling region of Oz (as described in *The Emerald City of Oz*) the city of Fuddlecumjig is inhabited by a whimsical race of people called the Fuddles. Each Fuddle is made of hundreds of fantastically shaped pieces of painted wood that fit together like a three-dimensional jigsaw puzzle. Whenever a visitor approaches, a Fuddle clatters into a heap of disconnected parts so that the visitor will have the fun of fitting him or her together again.

"Those are certainly strange people," Dorothy's Aunt Em said when she met the Fuddles, "but I really can't see what use they are, at all."

"Why, they amused us all for several hours," responded the Wizard. "That is being of use to us, I'm sure."

***Figure 125*** *Berrocal's Il Cavallo*

"I think they're more fun than playing solitaire or mumbletypeg," added Uncle Henry. "For my part, I'm glad we visited the Fuddles."


## ADDENDUM

For a full description of the 119,979 possible pieces for the burr puzzle, see Cutler's article cited in the Bibliography. For information about his computer programs for the burr and various puzzles that can be made with sets of six pieces, you can write to Cutler at his present address, 405 Balsam Lane, Palantine, IL 60067. A particularly fiendish version, which he calls Bill's Baffling Burr, is described by A. K. Dewdney in his *Scientific American* column, also cited in the Bibliography. It requires four shifts of pieces before the key piece can be removed. Dewdney challenged his readers to construct a similar puzzle in two dimensions and printed the best one received in his January 1986 column.

In 1980 Pentangle, the British puzzle-making firm, introduced The Chinese Cross Puzzle, designed by C. A. Cross. It is a boxed set of 42 mahogany pieces with instructions for making 314 different six-piece burr puzzles.


## BIBLIOGRAPHY

"Sculpture: Take Apart and Look Again." *Time*, May 23, 1969, p. 78.
*La Sculpture de Berrocal.* Giuseppe Marchiori. Brussels: La Connaissance Bruxelles, 1973.

### On burr puzzles

"The Six-Piece Burr." William Cutler, in *Journal of Recreational Mathematics*, 10, 1977–1978, pp. 241–250.
*Creative Puzzles of the World.* Peter Van Delft and Jack Botermans. Abrams, 1978, pp. 66–80.
"Computer Recreations." A. K. Dewdney, in *Scientific American*, October 1985, pp. 16–22.
*Puzzle Craft.* Stewart Coffin. Privately printed, 1985, pp. 32–60.
*Puzzles Old and New.* Jerry Slocum and Jack Botermans. University of Washington Press, 1986, pp. 68–87.
*Holey 6-Piece Burr!* William Cutler. Privately printed, 1986.
*A Computer Analysis of All 6-Piece Burrs.* Bill Cutler, privately printed, 1994.

# CHAPTER 19

# Sicherman Dice, the Kruskal Count and Other Curiosities

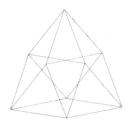

The equipment used in familiar games—dice, chessmen, checkers, cards and so on—has always been a rich source of problems in recreational mathematics. Here are some recent examples of such problems, chosen for their variety and elegance. Most of the questions are answered here, but I am withholding one solution for the answer section.

I shall begin with dice. Is it possible to number the faces of a pair of cubes in a way completely different from that of standard dice so that the cubes can be used in any dice game and all the odds will be exactly the same as they are when standard dice are used?

As far as I know, Col. George Sicherman of Buffalo was the first to pose and solve this question. The answer is yes, and the weird pair of dice illustrated in Figure 126 show the only way it can be done if we assume that each face must bear a positive integer. It does not matter how the six numbers are arranged on each die. Sicherman has placed

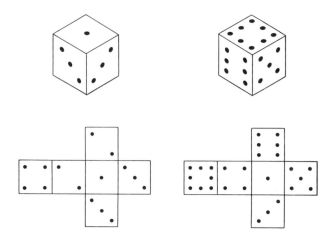

**Figure 126**  Sicherman dice

the numbers so that opposite sides of the left die total five and opposite sides of the right die total nine.

At the left in Figure 127 is the familiar matrix that displays all the ways each sum from 2 through 12 can be made with a pair of standard dice. There are 36 combinations. To determine the probability of throwing a given sum, *n*, count the number of *n*'s on the chart and divide by 36. For example, there are three 4's, and so the probability of throwing a 4 is 3/36, or 1/12. The corresponding chart for Sicherman dice is at the right in the illustration. It proves that the odds for each sum are exactly

|  | • | •. | •.. | •: | •:. | :: |
|---|---|---|---|---|---|---|
| • | 2 | 3 | 4 | 5 | 6 | 7 |
| •. | 3 | 4 | 5 | 6 | 7 | 8 |
| •.. | 4 | 5 | 6 | 7 | 8 | 9 |
| •: | 5 | 6 | 7 | 8 | 9 | 10 |
| •:. | 6 | 7 | 8 | 9 | 10 | 11 |
| :: | 7 | 8 | 9 | 10 | 11 | 12 |

|  | • | •. | •.. | •.. | •.. | :: |
|---|---|---|---|---|---|---|
| • | 2 | 3 | 3 | 4 | 4 | 5 |
| •.. | 4 | 5 | 5 | 6 | 6 | 7 |
| :: | 5 | 6 | 6 | 7 | 7 | 8 |
| •:. | 6 | 7 | 7 | 8 | 8 | 9 |
| :: | 7 | 8 | 8 | 9 | 9 | 10 |
| ::: | 9 | 10 | 10 | 11 | 11 | 12 |

**Figure 127**  *Chart of odds for standard dice* (left) *and Sicherman dice* (right)

the same as with ordinary dice. A casino could use Sicherman dice on the crap table without altering any of its betting rules or changing the vigorish (house percentage), although it might be hard to convince customers that the probabilities were unaltered.

To demonstrate that there is no other way to construct such a chart with positive integers takes a bit of doing that I shall not go into here. Sicherman also found that it is impossible to redesign a set of three or more dice that have the same odds as ordinary dice without utilizing his two dice. For example, Sicherman's pair plus one conventional die have the same odds as three ordinary dice, two pairs of Sicherman dice behave like four ordinary dice and so on.

The standard way numbers are arranged on Western dice (opposite sides total 7, and 1, 2 and 3 go counterclockwise around a corner) is involved in many puzzles and magic tricks and even in bits of numerology. The 4 edges of each face of a die represent the 4 seasons. The 12 edges of the cube stand for the 12 months. There are three pairs of digits on a die that add up to 7, the number of days in a week. If $7 \times 7 \times 7$ is added to $7 + 7 + 7$, the sum is $343 + 21$, or 364. Adding 1 for the die gives 365, the number of days in a year.

A die can be held so that one, two or three faces are visible. Is it possible to turn the die in different ways so that what is seen adds up to every number from 1 through 15? Curiously, only the unlucky 13 is impossible. Multiplying 13 by 4 (for the four corners of a face) gives 52, the number of weeks in a year and the number of cards in a deck.

Karl Fulves, a New Jersey magician, writer and computer scientist, recently invented an unusual extrasensory-perception trick based on the way the digits are arranged on a die. The magician hands someone a die, turns his back and gives the following instructions. The subject is asked to place the die on a table with any face uppermost. If the top number is even, he must give the die a quarter turn to the east (to his right). If the top number is odd, he must give the die a quarter turn to the north (away from him). This procedure is repeated sequentially, always obeying the rule: Turn east if the top face is even, turn north if it is odd. Every time the subject moves the die he calls out, "Turn." He does not, of course, reveal the number he started with or any subsequent top number.

After a few turns the magician tells the subject to stop as soon as 1 appears on top. The subject is then asked to give the die one additional turn (in compliance with the rule) and next to concentrate on the number this brings uppermost. It seems as though the number could be any one of four possibilities. Nevertheless, with his back still turned, the magician names the number.

A little experimentation discloses the secret. After at the most three turns the die enters the following loop: $1-4-5-6-3-2$, $1-4-5-6-3-2$. . . . Therefore the number following 1 is always 4. After three moves it is always safe to tell the subject to stop when 1 appears on top. It is safe after just two moves, except when the starting position is 6 up and 5 facing the subject.

The following combinatorial problem was sent to me several years ago by Christer Lindstedt, a correspondent in Sweden. Imagine a three-by-three-by-three cube formed with 27 standard dice. There are 54 pairs of face-to-face numbers in such a configuration. Multiply each pair of numbers and then add the 54 products. What is the minimum sum that can be achieved by a suitable arrangement of the dice? What is the maximum sum? I do not know the answer to either question, and I do not see any good way to find the answers without a computer. I am not even sure of the maximum and minimum sums for a cube of eight dice. The best results I have obtained are 306 and 40.

Here is a little-known chess task problem with a special case that provides a pretty puzzle, although the general case is unsolved. Can five queens of one color and three of another color be placed on a five-by-five chessboard so that no queen of one color attacks a queen of the other color? You can sketch the board on paper and use pawns or coins for queens. Surprisingly, there is only one solution (not counting rotations and reflections).

On the same order-5 board (a board with five cells on a side) it is impossible to place five queens so that more than three cells are unattacked or to place three queens so that more than five cells are unattacked. This fact suggests the more general problem. Given a board of order $n$ and $k$ queens of one color, what is the maximum number of unattacked cells that can be produced by a suitable arrangement? Of course, queens of a different color can be placed on the unattacked cells, so that this problem is the same as asking for the maximum number of, say, white queens that can be placed along with $k$ black queens on an order-$n$ board so that no queen of one color attacks a queen of the other color.

The general problem is meaningless for boards of orders 1 and 2, and it is easy to see that on the order-3 board a queen can be put on a corner or side cell to leave a maximum of two cells unattacked. The problem starts to get interesting when $n$ equals 4. It is not known whether there are unique patterns for $k$ queens on boards of order higher than 5, and finding a formula for the general problem seems difficult, if not impossible.

There are dozens of classic chess task problems involving knights. Some of them are given in Chapter 14 of my *Mathematical Magic Show*. Scott Kim has proposed the following knight task, which I have not seen before. Can 16 knights be placed on the standard chessboard so that each knight attacks just 4 others? Figure 128 displays the beautifully symmetric solution. The black lines, which show all the attacks, form a planar projection of the skeleton of the hypercube.

In 1977, Jan Mycielski, a mathematician at the University of Colorado, wrote to ask if the following problem, suggested by his colleague Richard J. Laver, is new. Can a finite set of equal-sized squares be drawn on the plane in such a way that every corner of every square is also a corner of at least one other square? The squares may overlap. Mycielski had found a set of 576 squares that solves the problem, and he wondered if that number could be reduced. Shortly thereafter he reported that another colleague had found a solution with 40 squares, and then that two other colleagues had independently brought the number down to 12 (see Figure 129). Finally, Andrzej Ehrenfeucht, a professor of computer science at the university, found a solution with eight squares.

I mentioned the problem to Kim without telling him of any of the above results. He staggered me by saying instantly, "It can be done with eight." He had, of course, remembered his 16-knight problem that solves Laver's problem with eight squares. There is surely no solution less than eight, but I have no proof of that. In three dimensions the problem can be solved with three squares. On the plane six identical equilateral triangles can be arranged so that every corner belongs to two triangles but no edge belongs to two triangles (see Figure 130). The solution is a

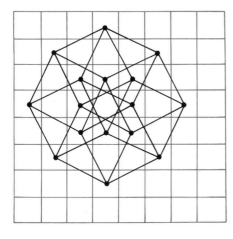

**Figure 128**
*Hypercube solution to Kim's knight problem*

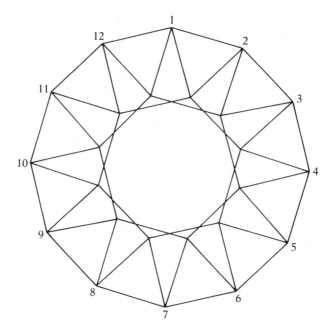

**Figure 129** *A 12-square solution to Richard J. Laver's problem*

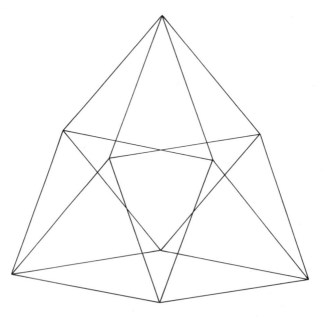

**Figure 130** *Projection of a four-dimensional polytope that solves a triangle problem*

planar projection of a four-dimensional polytope called a "triangular double prism."

David L. Silverman, author of *Your Move* (a splendid collection of puzzles based on games), invented a novel board game, not included in his book, involving a nonstandard chess piece that is usually called the amazon. An amazon combines the power of a queen and a knight. The game is played on a standard chessboard with two amazons and a supply of counters. Queens can be employed as amazons, but it is important to remember that each such piece also has the power of a knight.

White opens the game by placing his amazon on any cell. Black then places his amazon on any unattacked cell. From that point on the players take turns, each player transferring his amazon to any vacant cell not under attack by the other amazon. An amazon is not moved like a queen or a knight. It is simply picked up and placed on any cell that is not threatened. After each move a counter, say a penny, is placed on the vacated cell. A cell with a counter is out of play (henceforth no amazon can occupy it), but the counter does not block any attack. As the game proceeds the cells slowly fill with pennies until eventually a player is unable to find a safe spot for his amazon. The last player to move wins.

If Silverman's amazon game is played on an order-5 board, the first player wins immediately by occupying the center square. Since all cells are attacked, the second player cannot even put his amazon on the board. On the standard chessboard the second player can always win by Silverman's clever pairing strategy. He mentally divides the board into four eight-by-two rectangles and numbers the cells of each rectangle as is shown in Figure 131. (Each number from 1 to 8 appears twice in each

| 1 | 2 | 3 | 4 | 5 | 6 | 7 | 8 |
|---|---|---|---|---|---|---|---|
| 5 | 6 | 7 | 8 | 1 | 2 | 3 | 4 |
| 1 | 2 | 3 | 4 | 5 | 6 | 7 | 8 |
| 5 | 6 | 7 | 8 | 1 | 2 | 3 | 4 |
| 1 | 2 | 3 | 4 | 5 | 6 | 7 | 8 |
| 5 | 6 | 7 | 8 | 1 | 2 | 3 | 4 |
| 1 | 2 | 3 | 4 | 5 | 6 | 7 | 8 |
| 5 | 6 | 7 | 8 | 1 | 2 | 3 | 4 |

**Figure 131** *A pairing strategy for the amazon game*

rectangle.) After each play by White, Black simply occupies the cell that is in the same rectangle as, and that has the same number as, the cell occupied by White. The game is an excellent example of the extraordinary power of a trivial pairing strategy to win a game that appears to be quite difficult to analyze. The pairing strategy obviously applies to any board of an even order higher than 4, and it is easy to devise slightly more complicated pairing patterns for first-player wins on all boards of odd order higher than 5.

Ross Honsberger's *Mathematical Gems II* (Mathematical Association of America, 1976), as exciting a collection as his earlier volume, discloses for the first time a remarkable result in checker jumping that was discovered by the University of Cambridge mathematician John Horton Conway. Imagine a standard checkerboard divided in half as is shown in Figure 132. The bottom half is shaded and the rows of the unshaded half are numbered (from bottom to top) 1 through 4. Now imagine a fifth row just beyond the top edge. If a checker jumps off this edge, it is considered to have jumped to row 5.

All jumps must be orthogonal, that is, horizontal and vertical but not diagonal. As in checkers, pieces that have been jumped over are removed. The problem is to determine the minimum number of checkers that can be placed on the shaded half of the board in such a way that a sequence of jumps will place a checker on row $n$.

It is obvious that two is the minimum number of checkers required for getting a piece to row 1. They are placed as is shown at the top left in Figure 133, and one jump does it. Four is the minimum number of

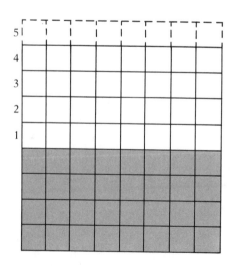

**Figure 132**
*John Horton Conway's
checker problem*

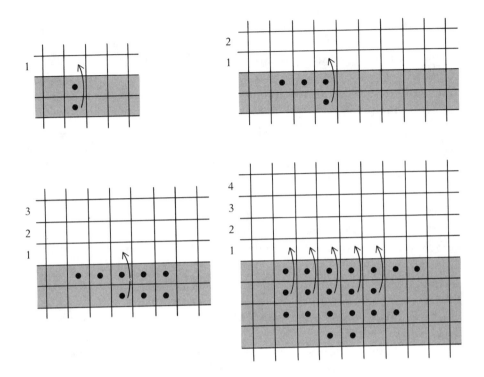

**Figure 133**   *How to get a checker to rows 1, 2, 3 and 4*

checkers needed to get to row 2. They can be placed as is shown at the top right in the illustration. The bottom checker jumps to row 1, and then the checker farthest to the left makes two jumps to end on row 2. To get to row 3 eight checkers must be placed in the starting position shown at the bottom left in the illustration. So far the minimum numbers are in a doubling sequence, but to get to row 4 the sequence is broken. At least 20 checkers are required. They can be arranged as is shown at the bottom right in the illustration. The arrows show how the jumps begin, and it should not be difficult to find a way to continue that will get a checker to row 4.

How many checkers are needed to get to row 5, that is, to jump one checker off the board? Astonishingly, no arrangement of checkers on the shaded cells will get a checker to row 5. The situation is even more hopeless. No matter how far the shaded section is extended downward and to the left and right, no pattern of checkers, however large, will boost a piece to row 5. Interested readers will find Conway's ingenious impossibility proof given in detail in Chapter 3 of Honsberger's book.

Turning to playing cards, there are so many new puzzles and magic tricks based on mathematical principles that it is agonizing to have space for only one. A few years ago Martin D. Kruskal, a physicist at Princeton University, made a strange discovery that is now known among card magicians as Kruskal's principle or the Kruskal count.

The principle is best explained by describing the card trick to which Kruskal first applied it. The subject shuffles a deck of cards and then thinks of any number from 1 to 10. He deals the cards slowly from the top, placing each card face up in a pile. As he deals, he counts to himself, noting the value of the card dealt at the chosen number.

Assume that he thought of 7 and the seventh card is a 5. Without hesitating in his deal he mentally calls the next card 1, and as he deals, he counts silently from 1 to 5. Suppose the fifth card is a 10. As before, the next card is mentally called 1, and as he deals, he counts silently from 1 to 10. This procedure is repeated until all 52 cards have been dealt. The cards at the end of each count, which determine how high the next count goes, are called "key cards." The last key card of the counting chain must be remembered by the magician's subject. It is the "chosen card" that has been selected by this randomized counting procedure.

It is unlikely, of course, that the final count will end on the last card. It is more likely that it will not be possible to complete the final count. The subject is cautioned to deal the cards slowly, in a regular rhythm, so that no hesitations in dealing will give away the key cards. If the final count cannot be completed, he still must remember the last key card, but in order not to reveal it he must continue dealing to the end.

To make the counting easier, the magician explains, all face cards are given a value of 5. Thus if a count ends on, say, a queen, the next count is not to 12 but to 5. To make the procedure clear Figure 134 shows a typical chain with the values of all the key cards indicated. The subject began by thinking of 4. The chosen card, which he remembers, is the jack of hearts. It has a value of 5, but because there are only three cards after it, the last count cannot be completed. It is obvious that this procedure, performed with a shuffled deck, can select any of the last 10 cards.

After the counting is finished and the subject has his chosen card in mind, the magician takes the deck, picks a card from the last 10 and places it face down on the table. The subject names his card. The magician's card is turned over, and it is *probably* the chosen card.

I have italicized "probably" to emphasize the strange way Kruskal's trick differs from almost all other tricks in which a magician finds a selected card. In this trick the magician cannot know the card with certainty. The probability that he will know the card is about 5/6.

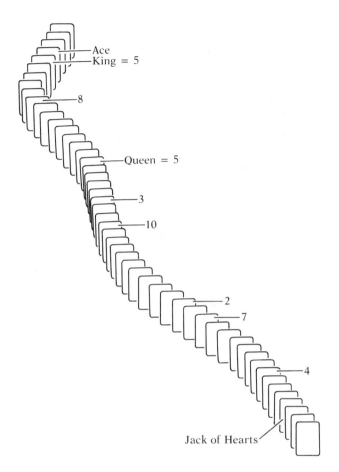

**Figure 134** *Typical chain for a trick based on Kruskal's principle*

Now for the curious secret. As the subject deals, the magician notes any fairly low card among the first few that are dealt. He treats this card as the first key of a chain that he counts silently to himself while the subject is counting his own chain. Kruskal's counterintuitive discovery is that about five times out of six the last key card of the magician's chain will be the same as the last card of the subject's chain! To put it another way, the probability is about 5/6 that any two arbitrarily started chains will intersect at a key card. Once this happens, the chains will be identical from that point on.

Giving face cards a value of 5 increases the number of keys in an average chain and thereby increases the probability of intersection. By starting his count on a low card among the first few (instead of picking an arbitrary number from 1 to 10), the magician adds slightly to the probable number of keys in his chain. This raises the odds of success a trifle more. If the trick is done with two decks shuffled together, a failure is extremely unlikely.

One of the best variations on Kruskal's trick comes from Cy Endfield of London. The trick is first performed as I have described it and is presented as a feat of telepathy. When the magician removes the (probably) chosen card, he notes the card immediately under it. If that card is low enough to allow another count, he continues the silent counting and remembers the last key card. The removed card is not returned to the deck.

The magician then hands the 51-card pack to the subject (without shuffling it) and asks him to repeat the trick, using a different starting number. "This time," the magician says, "I shall try to name your card by precognition." He writes a prediction on a sheet of paper that is folded and put aside. What he writes, of course, is the card he is remembering. Since the structure of the deck has not been altered, it is likely (with the same odds of about five to one) to be the second chosen card.

It should also be possible to program a computer to play the role of the psychic. Fifty-two punched cards would bear the names of the playing cards. The subject shuffles, selects a card by the Kruskal count and feeds the deck to the computer. The computer is programmed to guess the card and simultaneously to remember the card most likely to be chosen when the trick is repeated. If the first guess is correct, the chosen card is removed from the computer's deck and the trick is done again with a new starting number. This time the computer prints out the name of the card without examining the deck.

Even a computer will not always be right, but the fact that the trick sometimes fails makes it all the more impressive. When Uri Geller failed on the Johnny Carson show a few years ago (Carson, a former magician, guarded the test materials carefully before going on camera), Merv Griffin declared that the failure proved to him that Geller's powers were genuine. Magician's tricks, Griffin explained, always work. We all know how the "force" comes and goes. Why should a psychic robot control it any better than a humanoid?

## ANSWERS

The eight-dice problem was to put them together to form a cube so that the sum of the 12 products of the 12 pairs of touching faces is (1) minimum and (2) maximum. My figures of 40 and 306 were correct.

William Funkenbush pointed out that if four of the dice are mirror images of standard dice (as they are in Japan), the minimum remains 40 but the maximum rises to 308. Minimum and maximum solutions are shown in Figure 135.

The same problem with 27 dice is answered with a minimum of 294 and a maximum of 1028. Solutions are shown in Figure 136. The minimum is unique except for two possible orientations of the central die. Kenneth Jackman was the first to solve this difficult task, followed by Leonard Lopow, David Vanderschel, Alan Cuthberton and Paul Stevens. Lopow and Vanderschel solved the problem without computer aid.

The chess task on the order-5 board has the unique solution (excluding rotations and reflections, of course) shown in Figure 137. I had given this problem in slightly different form in a 1972 column on chess tasks that is reprinted in my *Wheels, Life, and Other Mathematical Amusements*. In that chapter's addendum I discuss the general problem of placing $k$ queens on order-$n$ boards to maximize the number of unat-

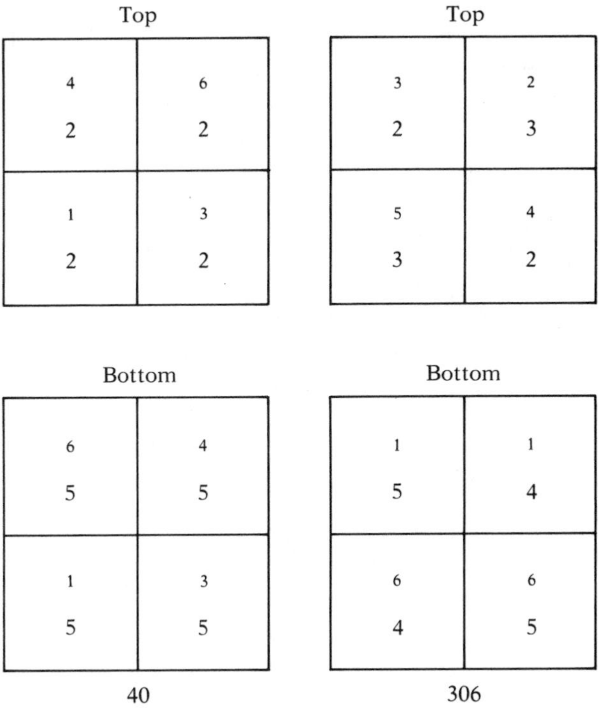

**Figure 135** *Minimum (40) and maximum (306) solutions to the eight-dice problem. Large number is top face, small number is the side facing north.*

| 6 5 | 5 6 | 4 5 | | | 2 1 | 2 1 | 3 1 |
|-----|-----|-----|---|---|-----|-----|-----|
| 5 4 | 2 6 | 3 6 | **Top** | | 4 1 | 2 1 | 2 3 |
| 3 5 | 1 4 | 1 5 | | | 4 1 | 6 2 | 6 3 |

| 5 4 | 6 5 | 6 4 | | | 2 4 | 1 4 | 1 4 |
|-----|-----|-----|---|---|-----|-----|-----|
| 3 1 | 6 4 | 4 5 | **Middle** | | 2 4 | 2 6 | 4 2 |
| 1 4 | 2 1 | 2 4 | | | 6 4 | 6 2 | 6 3 |

| 4 2 | 6 3 | 6 2 | | | 1 5 | 1 5 | 3 5 |
|-----|-----|-----|---|---|-----|-----|-----|
| 3 1 | 4 1 | 5 3 | **Bottom** | | 4 5 | 5 4 | 2 6 |
| 1 2 | 2 1 | 3 2 | | | 4 5 | 4 6 | 6 5 |

294                    1028

**Figure 136** *Minimum (294) and maximum (1028) solutions to the 27-dice problem. Large number is the top face, small number is the side facing north.*

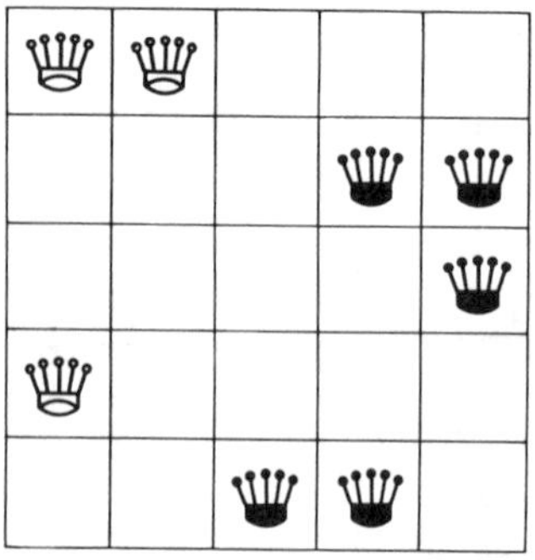

**Figure 137** *Solution to chess task*

tacked cells. The general problem remains unsolved, but Ronald Graham and Fan Chung at Bell Labs are continuing to work on it with surprising results that they hope to publish eventually.

## ADDENDUM

I said that Sicherman dice could be used in any dice game without altering the odds, but this is not true. Gary Goodman pointed out that in backgammon, for example, doubles are treated as special cases. Odds for Sicherman-dice doubles are not the same as for standard dice; it is not even possible to throw a double 2. William Funkenbush wrote to say that even at casino crap tables Sicherman dice cannot be used for side bets such as making a certain sum "the hard way"—that is, with identical numbers on both dice. Two hard-way bets, 4 and 10, obviously cannot be made at all.

Many readers sent algebraic proofs of the uniqueness of Sicherman dice, all essentially the same as Sicherman's proof—by way of generating polynomials and their unique factorization. Two published papers generalized Sicherman's discovery to $k$ dice, each with $m$ faces. The generalized dice can be modeled with Platonic solids, rolling cylinders with $m$ faces, spinners with $m$ numbers or urns with $m$ balls. See "Renumbering of the Faces of Dice" by Duane Broline in *Mathematics Magazine* (vol. 52, 1979, pages 312–315) and "Cyclotomic Polynomials and Nonstandard Dice" by Joseph Gallian in *Discrete Mathematics* (vol. 27, 1979, pages 245–259). For a later rediscovery of essentially the same results, see "Dice with Fair Sums" by Lewis Robertson, Rae Shortt and Stephen Landry in *American Mathematical Monthly* (vol. 95, 1988, pages 316–328).

Can a pair of dice be renumbered with positive integers so that every possible sum is tossed with equal probability? This, too, has only one solution. See the chapter on dice in my *Mathematical Magic Show*. Karl Fulves's dice trick is in his privately printed Faro Possibilities (1970, pages 33–34), where it is followed by a similar but more complicated die trick involving a cycle of 24 steps.

I said I knew of no proof that eight squares are minimal for the problem of placing squares so every corner of each square is a corner of at least one other square. Karl Scherer of West Germany provided a lengthy proof. The problem generalizes to the placing of $n$-sided polygons in any number of dimensions, and several readers sent letters about it. The generalization plunges into the fascinating structures of polytopes in higher spaces. The polytope whose projection solves the triangle-placing problem belongs to a class called "Hessian polytopes." They are discussed in H. S. M. Coxeter's classic work, *Regular Complex Polytopes*, now happily available in a Dover paperback edition.

Conway's checker theorem was generalized to 3-space by Eugene Levine and Ira Papick in "Checker Jumping in Three Dimensions" in *Mathematics Magazine* (vol. 52, September 1979, pages 227–231). The authors show that a checker cannot jump higher than level seven on the cubical board.

Card magicians interested in tricks based on Kruskal's principle will find articles on them in two magic periodicals: "The Kruskal Principle" by Karl Fulves and myself in *The Pallbearers Review* (June 1975) and Charles Hudson's "Card Corner" column in *The Linking Ring* (December 1976, pages 82–87), containing a variety of ideas from the Chicago card expert Ed Marlo. The same column (December 1957 and March 1978) discusses related card tricks based on what is called the "Kraus principle."

Rereading "The Disappearance of Lady Frances Carfax," a Sherlock Holmes tale in the book *His Last Bow*, I was startled to encounter a remark of Holmes that seems to anticipate the Kruskal count: "When you follow two separate chains of thought, Watson, you will find some point of intersection which should approximate to the truth."

# CHAPTER 20

## Raymond Smullyan's Logic Puzzles

"I now introduce Professor Smullyan, who will prove to you that either he doesn't exist or you don't exist, but you won't know which."

—MELVIN FITTING, *introducing Raymond Smullyan to an undergraduate mathematics club*

Raymond M. Smullyan's *What Is the Name of This Book?* (Prentice Hall, 1978) is the most original, most profound and most humorous collection of problems in recreational logic ever written. It contains more than 200 brand-new puzzles, all concocted by the ingenious author and interspersed with mathematical jokes, lively anecdotes and mind-bending paradoxes. The book culminates with a remarkable series of story problems that lead the reader into the core of the late Kurt Gödel's revolutionary work on undecidability.

Who is Smullyan? He was born in 1919 in New York, studied philosophy under Rudolf Carnap at the University of Chicago and received his Ph.D. in mathematics at Princeton University. He is currently a professor of philosophy at Indiana University in Bloomington and professor emeritus of mathematics and philosophy at Lehman College and CUNY Graduate Center in New York City.

Among experts on logic, recursion theory, proof theory and artificial intelligence he is best known as the author of two elegant little treatises: *First-Order Logic* (Springer-Verlag, 1968) and *Theory of Formal Systems* (Princeton University Press, 1961). His article in *The Encyclopedia of Philosophy* on Georg Cantor's famous continuum problem is a marvel of lucid compression. In 1977 Harper & Row published his first nontechnical work, *The Tao Is Silent*, one of the best introductions to Taoism I have seen.

Smullyan's main hobbies are music (he is an accomplished classical pianist), magic (in his youth he was a part-time professional) and chess. He has recently finished work on two collections of remarkable chess problems, *Chess Mysteries of Sherlock Holmes* and *Chess Mysteries of the Arabian Knights*, in which each problem is embedded in an appropriate pastiche. Add an enticing literary style, a huge sense of humor and a Carrollian love of paradox and you have the flavor of *What Is the Name of This Book?*

The book opens with a true story that introduces one of Smullyan's central themes. On April Fool's Day, when Smullyan was six, his older brother Émile told him he was going to be fooled as he had never been fooled before. All day long Smullyan waited for the prank, and he lay awake that night still waiting for it. Finally Émile revealed the joke: Raymond had expected to be fooled and so, by not doing anything, Émile had fooled him.

"I recall lying in bed long after the lights were turned out." Smullyan writes, "wondering whether or not I had really been fooled. On the one hand, if I wasn't fooled, then I did not get what I expected, hence I was fooled. . . . But with equal reason it can be said that if I was fooled, then I *did* get what I expected, so then, in what sense was I fooled?. . . . I shall not answer this puzzle now; we shall return to it in one form or another several times in the course of this book."

After an introductory section on some classic brainteasers, many with amusing new twists, Smullyan introduces three types of people who will participate in most of the problems that follow: "knights," who always tell the truth; "knaves," who always lie; and "normals," who sometimes tell the truth and sometimes lie.

It is amazing how much can be deduced from just a few lines of dialogue with these individuals. For example, on an island inhabited only by knights and knaves Smullyan comes on two men resting under a tree. He asks one of them: "Is either of you a knight?" When the man—

call him *A* — responds, Smullyan instantly knows the answer to his question. Is *A* a knight or a knave? What is the other man?

It seems as though there is not enough information to solve this problem, but the key lies in the fact that *A*'s answer enabled Smullyan to find the solution. If the answer had been yes, Smullyan would have learned nothing. (If *A* is a knight, one or both of the men could be knights, and if *A* is a knave, both of them could be knaves.) Therefore *A* must have said no. Now, if *A* were a knight, he would have had to say yes, but since he said no he must be a knave. Because he is a knave his no is false; therefore at least one knight must be present. Hence the other man is a knight.

Lewis Carroll's Alice soon enters the book. We find her wandering around in the Forest of Forgetfulness, where she is unable to remember the day of the week (see Chapter 3 of *Through the Looking-Glass*). In the forest she meets the Lion and the Unicorn. The Lion lies on Monday, Tuesday and Wednesday and the Unicorn lies on Thursday, Friday and Saturday. At all other times both animals tell the truth. "Yesterday was one of my lying days," says the Lion. "Yesterday was one of my lying days too," says the Unicorn. Alice, who is as smart as Smullyan, is able to deduce the day of the week. What is it?

Enter more *Looking-Glass* characters: the Tweedle brothers, the White King, Humpty Dumpty and the Jabberwock. Tweedledee behaves like the Unicorn, Tweedledum like the Lion. After Alice has solved a number of problems based on conversation with the Tweedles, Humpty Dumpty discloses a well-kept *Looking-Glass* secret: There is a third brother, identical in appearance with Dum and Dee, named Tweedledoo. Doo always lies. Alice is now dreadfully upset because all her previous deductions may be wrong. On the other hand, Humpty may be lying, and Tweedledoo may not exist. Four accounts are given of what happens next, and the reader is asked to deduce which one is correct and whether or not Tweedledoo exists.

The scene then shifts to Shakespeare's *The Merchant of Venice* and that famous puzzle occasion when Portia presents her suitor with three caskets, gold, silver and lead, each of which bears a different inscription. Only one casket contains Portia's portrait, and if the suitor guesses it correctly, Portia will marry him. (It is not generally realized, by the way, that Portia gives her suitor a whopping hint by singing, "Tell me where is fancy bred. / Or in the heart or in the head?" The last words of both lines rhyme with lead, the correct choice.)

By varying the inscriptions on the caskets Smullyan creates a series of remarkable problems that lead the reader closer and closer to Gödel's discovery. The first problem is shown in Figure 138. Portia, who never lies, explains to her suitor that at most one inscription is true. Which casket should he choose?

Smullyan improvises clever variations on the Portia theme. In some problems each casket bears two inscriptions. We also learn that there are two casket makers: Bellini, who always puts true inscriptions on his caskets, and Cellini, who always puts on false ones.

Consider another casket problem, shown in Figure 139. The gold casket bears the inscription "The portrait is not in here" and the silver casket bears the inscription "Exactly one of these two statements is true." The two inscriptions present a logical dilemma of enormous importance in the history of modern semantics. The suitor reasons: If the silver statement is true, then the gold one is false. If the silver statement is false, then the inscriptions are either both true or both false. They cannot both be true if the silver statement is false, and so they are both false. In either case the gold statement is false; therefore the gold casket must contain the portrait. The suitor triumphantly opens the gold casket but finds to his horror that it is empty. The portrait is in the other casket. What is wrong with his reasoning?

The error he makes is in assuming that the statement on the silver casket is either true or false. This problem involves us in the modern concept of metalanguages. It is only permissible to discuss the truth values of a particular language in a larger language, or metalanguage, that contains the first language as a subset of its terms. When a language refers to its own truth values, the result is often a logical contradiction. Without metastatements about the truth or falsity of the casket inscrip-

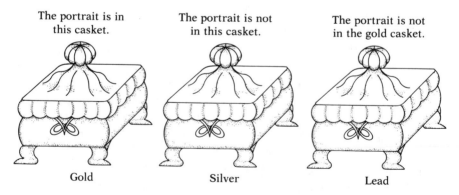

|                           |                               |                                |
|---------------------------|-------------------------------|--------------------------------|
| The portrait is in this casket. | The portrait is not in this casket. | The portrait is not in the gold casket. |
| Gold                      | Silver                        | Lead                           |

*Figure 138  Portia's first casket test*

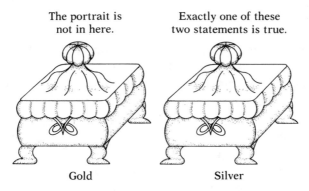

The portrait is not in here.

Exactly one of these two statements is true.

Gold          Silver

**Figure 139**   *Where is Portia's portrait?*

tions or information about how their truth values are related the inscriptions can be meaningless.

Inspector Leslie Craig of Scotland Yard now strides into the book, and Smullyan gives us a variety of mysteries from the inspector's files that can be solved by careful logical deduction. The first one is the simplest:

> An enormous amount of loot had been stolen from a store. The criminal (or criminals) took the heist away in a car. Three well-known criminals, *A*, *B* and *C*, were brought to Scotland Yard for questioning. The following facts were ascertained:
>
> **(1).**  No one other than *A*, *B* and *C* was involved in the robbery.
>
> **(2).**  *C* never pulls a job without using *A* (and possibly others) as an accomplice.
>
> **(3).**  *B* does not know how to drive.
>
> Is *A* innocent or guilty?

In the pages that follow Smullyan is concerned with such practical tasks as how to avoid werewolves, how to choose a bride, how to defend yourself in court and how to marry a king's daughter. For instance, suppose you want to convince a prospective bride who is unusually fond of knaves that you are a rich knave. (You must be rich or poor.) Can it be done with a single sentence? Yes. You have only to say, "I am a poor knave." The girl knows at once that you cannot be a knight because a knight would not lie and say he is a poor knave. Since you are a knave your statement must be false; therefore you are a rich knave. Suppose the girl is attracted only to knights. What sentence will convince her you are a rich knight?

The next section introduces logic puzzles based on the conditional statement of the form "If $P$ is true, then $Q$ is true." The two statements are connected by the relation of implication, an understanding of which is absolutely indispensable in understanding the propositional calculus. Smullyan plays with the familiar paradoxes of implication and then gives 18 ingenious puzzles no reader can think through without acquiring a firm grasp on the logical principles involved.

The next setting is the Island of Baal, the only place on the earth where someone knows the answer to the superultimate metaphysical question: "Why is there anything at all?" The island is inhabited only by knights and knaves. After a series of encounters with the local folk Smullyan proves that the Island of Baal cannot exist.

The proof of nonexistence does not apply to the next island Smullyan visits: the Island of Zombies. Here there is no easy way to distinguish the zombies, who always lie, from the human beings, who always tell the truth. Life is further complicated by the fact that all yes-or-no questions are answered with either "bal" or "da," but we do not know which means yes and which means no. Suppose you ask a native whether "bal" means yes and he answers "bal." You cannot determine what "bal" means, but can you tell whether the speaker is a human being or a zombie? Is it possible to find out what "bal" means in just one yes-or-no question?

Transylvania is equally confusing. Here the human beings (truth-tellers) do not look different from the vampires (liars) and half of the inhabitants are insane. The insane believe all true propositions are false and all false propositions are true. Thus there are four types of Transylvanians: sane human beings, insane human beings, sane vampires and insane vampires. Anything a sane human being says, of course, is true and everything an insane human being says is false. Conversely, anything a sane vampire says is false and anything an insane vampire says is true. Luckily all questions are answered in English. How can you determine in one yes-or-no question whether a Transylvanian is a vampire? How can you find out in one yes-or-no question whether he is sane?

Smullyan, eager to know whether Dracula is alive or dead, puts the question to various Transylvanians. The reader is asked to deduce the answer from the dialogue. This section culminates in a grand ball at Count Dracula's castle, where the complexities are compounded by the fact that all questions are answered with "bal" and "da," as on the Island of Zombies. As a result there are three variables to worry about: Is the speaker sane, is he human and what does "bal" mean? Smullyan eventually discovers that Dracula is alive but insane.

One chapter of the book is titled "How to Prove Anything." After examining a sophistry in Plato's dialogue *Euthydemus* in which one speaker proves that another speaker's father is a dog, Smullyan discusses a number of curious devices by which one can seemingly prove that anything—God, Satan, unicorns, Santa Claus and so on—exists. One of the devices is derived from the traditional ontological argument for God. Some of the others are variants of a subtle method discovered by the mathematical logician J. Barkley Rosser.

For example, consider the following sentence: "If this sentence is true, then Santa Claus exists." Smullyan writes: "If the sentence is true, then surely Santa Claus exists (because if the sentence is true, then it must also be true that if the sentence is true, then Santa Claus exists, from which it follows that Santa Claus exists); hence what the sentence says is the case, so the sentence is true. Hence the sentence is true, and if the sentence is true, then Santa Claus exists. From this it follows that Santa Claus exists." The argument is unsound, but without an understanding of the role of metalanguages it is not easy to explain exactly why.

The penultimate chapter introduces the familiar "liar paradox" ("This statement is false") and its many disguises and variants. Smullyan presents some of the deepest paradoxes of logic and set theory in a way that makes them clearer than they have ever been made before. For instance, here is his explanation of the famous paradox, known as Richard's paradox, that is the basis for Gödel's undecidability proof.

A mathematician has a book called *The Book of Sets*. On each page he lists or describes a set of counting numbers. The pages are numbered consecutively. Can we describe a set of positive integers that cannot be listed in the book?

We can. If a number $n$ belongs to the set listed on page $n$, call it an extraordinary number. If $n$ does not belong to the set on page $n$, call it an ordinary number. Now consider the set of all ordinary numbers. Assume that this set is listed on a certain page. The number of the page cannot be ordinary because if it were, that number would be on the page and consequently it would be extraordinary. On the other hand, it cannot be extraordinary because in that case it would have to appear on the page and we assumed that the page lists only ordinary numbers. This contradiction forces us to abandon the assumption that the set of ordinary numbers can be listed. Therefore there is a set of positive integers that cannot be listed in the book.

Now we are ready for Smullyan's climactic chapter on Gödel's discovery; this final chapter is the best introduction I know to that great watershed in the history of the study of the foundations of mathematics. Since the days of Leibniz mathematicians have dreamed that someday all mathematics would be united in one vast system in which every statement that could be formulated could be proved true or false. Leibniz even extended the dream to philosophical disputes. "If controversies were to arise," he wrote, "there would be no more need of disputation between two philosophers than between two accountants. For it would suffice to take their pencils in their hands, to sit down to their slates, and to say to each other (with a friend to witness, if they liked): Let us calculate."

This dream was shattered forever by Gödel's paper of 1931. In the paper the 25-year-old Gödel showed that the deductive system of Alfred North Whitehead and Bertrand Russell's *Principia Mathematica*, as well as related systems such as standard set theory, contains undecidable statements, that is, statements that are true but cannot be proved true within the system. More precisely, Gödel showed that if a system like that of *Principia Mathematica* satisfies certain reasonable conditions such as consistency (freedom from contradiction), then it allows the formation of sentences that are undecidable. At the same time he showed that if such a system is consistent, there is no way to prove the consistency within the system.

These results apply to any deductive system rich enough to contain arithmetic. Even in ordinary arithmetic there are statements that are true but unprovable. (Very simple systems, such as arithmetic without multiplication, are free from any undecidable statements.) Moreover, it is not possible to prove the consistency of arithmetic within arithmetic.

One can, of course, enlarge arithmetic by adding new axioms so that the enlarged system allows proof of any formerly undecidable statement. Alas, the situation is as hopeless as it was before. It can be shown by Gödel's same arguments that the enlarged system will contain new undecidable statements and that the consistency of the enlarged system cannot be proved within it. This construction of increasingly larger systems can go on forever, but it will never reach a level where undecidable statements can be banished from a system or a consistency proof can be devised within the system.

There is a famous unsolved problem in arithmetic called Goldbach's conjecture. It asserts that every even number greater than 2 is the sum of two primes. No one has proved it or found a counterexample. It is possible that Goldbach's conjecture is Gödel-undecidable. If that is so, it means the conjecture is true but unprovable within arithmetic. It is true because if it were false, a counterexample would exist, and that would make the conjecture decidable.

An even more disturbing aspect of this situation is the lack of any constructive way of showing that number theorists will not someday find an arithmetical proof that Goldbach's conjecture is true and also a proof that it is false! Mathematicians hope and believe it will never happen to any arithmetical theorem because if it does, it would reduce arithmetic and all higher mathematics to a shambles. (It is easy to show that if a deductive system, in which proofs by contradiction are valid, contains even one contradiction, then it is possible to prove any statement whatsoever in the system.) Platonist mathematicians who regard the axioms of arithmetic as being true and the rules of reasoning as being correct have no such worries because they believe no contradictions can arise. The purely constructive formalists, however, have no such guarantees.

Gödel's undecidable statements are undecidable only within a given system. In 1936 papers by A. M. Turing and Alonzo Church established the existence of problems that are undecidable in a deeper sense. They proved the existence of problems for which there is no finite algorithm, or step-by-step procedure, that will solve them. Examples of these absolute undecidables include the famous halting problem of Turing-machine theory, the color-domino problem in tiling theory, problems in John Horton Conway's game of Life and many more. At no time in the future will it be possible in any logically consistent world to build a computer, however powerful, that by twiddling symbols will be able to solve such problems in a finite number of steps.

Since 1936 all kinds of ways of establishing Gödel's results and the related Church-Turing results have been devised, some of them simpler than Gödel's original method. Smullyan presents Gödel's proof in a delightful way by imagining a Gödel island inhabited only by knights and knaves. Knights who have "proved themselves" to be knights are called established knights. Knaves who have proved themselves to be knaves are called established knaves. The inhabitants have formed clubs for which the following conditions hold:

1. The set of all established knights forms a club.
2. The set of all established knaves forms a club.
3. Every club $C$ has its complement: a club consisting of all people on the island who are not in club $C$.
4. Given any club, there is at least one inhabitant who professes to be a member of that club.

Smullyan is now able to show in just three paragraphs of simple, nontechnical argument that there is at least one unestablished knight on the island and least one unestablished knave. If we regard the knights as being true sentences, the established knights as being provably true sentences, the knaves as being false sentences and the established knaves as being provably false sentences, the results of Smullyan's argument correspond to Gödel's results. In only three more sentences Smullyan establishes a related theorem of Alfred Tarski's, that neither the set of knaves nor the set of knights forms a club.

There is a well-known version of the old liar paradox where instead of a single sentence there are two sentences, one on each side of a card. One sentence states, "The sentence on the other side of this card is true." The other states, "The sentence on the other side of this card is false." Neither sentence refers to itself and yet the contradiction is obvious. Similarly, Smullyan imagines what he calls "doubly Gödelian islands" of knights and knaves that satisfy the following condition: Given any two clubs $C_1$ and $C_2$, there are inhabitants $A$ and $B$ such that $A$ claims that $B$ is a member of $C_1$ and $B$ claims that $A$ is a member of $C_2$. The study of such double islands is one of Smullyan's favorite hobbies. He discusses several of his discoveries about them and gives some new problems that have not yet been solved.

The book ends with a truly astonishing version of Gödel's construction of an unprovable sentence. Consider the following statement: "This sentence can never be proved." "If the sentence is false," writes Smullyan, "then it is false that it can never be proved, hence it *can* be proved, which means it must be true. So, if it is false, we have a contradiction, therefore it must be true.

"Now, I have just proved that the sentence is true. Since the sentence is true, then what it says is really the case, which means that it can never be proved. So how come I have just proved it?"

The fallacy, Smullyan explains, is that it has not been made clear just what is meant by provable. Consider a revised version of the sentence: "This sentence is not provable in system $S$." The paradox magically vanishes! "The interesting truth is that the above sentence must be a true sentence which is not provable in system $S$." (It is assumed, of course, that everything provable in $S$ is true.) In this form the sentence is "a crude formulation of Gödel's sentence $X$, which can be looked at as asserting its own unprovability, not in an absolute sense, but only within the given system."

At this point Smullyan suddenly remembers that he has not yet answered the question "What is the name of this book?" The name is. . . . But see the book's last page for the answer.

# ANSWERS

1.  The Lion can say "I lied yesterday" only on two days: Monday and Thursday. The Unicorn can make the same statement only on Thursday and Sunday. Therefore the only day on which both the Unicorn and the Lion can make the statement is Thursday.

2.  The inscriptions on the gold and lead caskets say the opposite, so that one of them must be true. Since at most only one statement is true, the statement on the silver casket is false. The portrait is therefore in the silver casket.

3.  If *B* is innocent, then we know (by fact 1) that either *A* or *C* is guilty. If *B* is guilty, he must have had an accomplice because he cannot drive; therefore again *A* or *C* must be guilty. Consequently *A* or *C* or both are guilty. If *C* is innocent, *A* must be guilty. If *C* is guilty, then (by fact 2) *A* is also guilty. Therefore *A* is guilty.

4.  You say "I am not a poor knight." The girl reasons that if you were a knave, you would indeed not be a poor knight; therefore your statement would be true. Since a knave never makes a true statement, the contradiction eliminates the assumption. Hence you are a knight. Knights speak truly, and so you are not a poor knight.

5.  An inhabitant on the Island of Zombies has replied "Bal" to the question, "Does bal mean yes?" If bal means yes, then bal is the truthful answer; therefore the speaker is human. If bal means no, then that too is truthful; therefore the speaker is human. It is not possible to determine what "Bal" means, but the answer does prove that the islander is human.

6.  To determine in one yes-no question what "Bal" means ask the islander if he is human. Since both human beings and zombies answer yes to such a question, if the islander answers, "Bal," the word means yes. If the islander answers "Da," then "Da" means yes and "Bal" means no.

7.  To tell whether a Transylvanian is a vampire by asking one yes-no question, ask him if he is sane. A vampire will say no and a human being will say yes. (I leave the proof to the reader.) To tell whether the Transylvanian is sane ask him if he is a vampire.

## ADDENDUM

Since I reviewed Smullyan's first book of logic problems, he has written other nontechnical books that you will find listed in the bibliography. I have also cited a perspicacious review of *What Is the Name of This Book?* by the eminent philosopher Willard Van Orman Quine and an entertaining article about Smullyan that appeared in *Smithsonian* magazine.

James Kennedy reminded me in a letter that a title of an earlier book, similar in self-reference to the title of Smullyan's book, was a book of magic tricks by the American magician Ted Annemann. He called it *The Book Without a Name.*

## BIBLIOGRAPHY

### Puzzle books by Raymond Smullyan

*The Chess Mysteries of Sherlock Holmes.* Knopf, 1979.
*The Chess Mysteries of the Arabian Knights.* Knopf, 1981.
*The Lady or the Tiger?* Knopf, 1982.
*Alice in Puzzle-Land.* Morrow, 1982.
*To Mock a Mockingbird.* Knopf, 1985.
*Forever Undecided: A Puzzle Guide to Gödel.* Knopf, 1987.

### Philosophical works by Smullyan

*The Tao Is Silent.* Harper & Row, 1977.
*This Book Needs No Title.* Prentice-Hall, 1980.
*Five Thousand B.C. and Other Philosophical Fantasies.* St. Martin's, 1983.
*Incompleteness and Undecidability.* Oxford University Press, forthcoming.

### About Smullyan

"Knights and Knaves." Willard Van Orman Quine, in the *New York Times Book Review*, May 28, 1978.
"The Puzzling and Paradoxical Worlds of Raymond Smullyan." Ira Mothner, in *Smithsonian*, 13, 1982, pp. 115–128.

### About Gödel and undecidability

*Computability* and *Unsolvability.* Martin Davis, McGraw-Hill, 1958.
*Gödel's Proof.* Ernest Nagel and James Newman. New York University Press, 1959.
*The Undecidable.* Martin Davis, ed. Raven Press, 1965.

# CHAPTER 21
∙ ∙ ∙ ∙ ∙ ∙

# The Return
# of Dr. Matrix

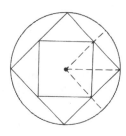

It's still the same old story.

—HERMAN HUPFELD *As Time Goes By**

In the last chapter of *The Magic Numbers of Dr. Matrix* (Prometheus, 1985), I reported the sad news of the death of my old friend, the famous numerologist Dr. Irving Joshua Matrix. I had last seen him and his half-Japanese daughter, Iva, in Istanbul in 1980. Disguised as an Arab named Abdul Abulbul Amir, Dr. Matrix was then on a top-secret mission for the CIA. When I returned to New York, I learned of his tragic rendezvous on the banks of the Danube, near Izmail, with a Russian KGB agent named Ivan Skavinsky Skavar. The two men, said the *New York Times*, apparently fired their revolvers simultaneously. Skavar's body was tossed by the Russians into the Black Sea. Amir was buried near Izmail in the Reichenbach Falls Cemetery.

---

Early in 1987 I attended an international conference on fractal geometry, held at the posh Ritz Hotel in Lisbon. During lunch with Benoît Mandelbrot, the father of fractals and the conference's honored guest and keynote lecturer, a loudspeaker on the wall summoned me to a telephone.

To my astonishment and delight, it was Iva! She refused to say how she knew I was in Lisbon, but she asked if it would be possible for me to visit her in Casablanca. After all, she said, Casablanca was only a short plane hop from Lisbon. She had, she added, some startling information to disclose about her father.

We arranged to meet for dinner at Rick's Place, Casablanca's well-known gambling casino, restaurant and bar. It is the same spot that was the setting for the 1942 motion picture *Casablanca* starring Ingrid Bergman, Humphrey Bogart, Claude Raines and Paul Henreid. As most aficionados of this classic cult film know, it was based on an unproduced play called "Everybody Comes to Rick's" — a play in turn based on real people. Readers may be interested to learn what happened to these people. In 1943 Ilsa's husband, Victor Laszlo, was assassinated by a revenging Nazi. A few years later Ilsa Lund and Richard Blaine were reunited in Paris, where they were married, with Rick's friend Captain Louis Renault as best man. They had one child, Richard, Jr.

After Richard senior died in 1957, Ilsa returned to Stockholm with her son. They lived there until her death in 1982. In 1983 Rick, Jr. bought the Café-Americain that his father had once owned, named it "Rick's Place" and has been managing it ever since.

Casablanca, a bustling port city of almost 3 million, was what I had expected — a city sharply bifurcated into regions of great wealth and abject poverty. Tall white office buildings with fashionable French shops gleamed in the bright African sun. Not far from El Mansour, my hotel, were the mean, dirty, gloomy streets of the poor, festering with crime, drugs, disease and misery. As my taxi rolled past one of the *Souks* (old markets), I could smell the spices and the smoke from kabobs cooking on open fires.

Rick's Place, just beyond the *Souk*, was noisy, smoky and jammed with European tourists and native Muslims, most of them speaking French. Rick himself, chain-smoking like his father, led me to the table where Iva smiled and waved. She had aged some but looked as stunning as ever. A tall white-haired man sat opposite her, his back to me. When I circled the table and saw his face, I almost collapsed. It was Dr. Matrix! His hawk-like visage was sombre as he stood up to buss me on both cheeks, but his emerald eyes, one of them behind a monocle, glinted with amusement.

I was totally at a loss for words while Iva filled me in on what had happened. The Russian agent died instantly from a bullet through his brain, but Abdul, or rather Dr. Matrix, had fallen unconscious after Ivan's bullet grazed his left temple. Agents from the CIA were nearby. Their helicopter carried Dr. Matrix to Morocco, where he quickly recovered and was given a new identity. Two peasants from Izmail were paid handsomely to testify that they had witnessed the deaths of both men. After a sham funeral, an empty casket was buried.

Dr. Matrix said little until Iva finished. He assured me that all danger of retaliation by the KGB had passed, but he was currently on another hush-hush assignment about which he could give no information beyond the fact that he was living in Casablanca as an expatriate poet from Indiana named Jasper Whitcomb Lundy. Iva, using a stage name she asked me not to divulge, was working as a belly dancer at a nightclub in the nearby resort town of Ain Diab. "It's my uncover," she said.

"Jasper Whitcomb Lundy," I repeated. "Why such an odd name?"

"No two letters are alike," said Dr. Matrix. "That's quite rare in such a long name. Observe also that its six vowels are in AEIOUY order."

"Your silver belt buckle," I remarked. "I see engraved on it the ancient Chinese *lo shu*, or magic square, with its odd digits on gray cells." (See Figure 140.)

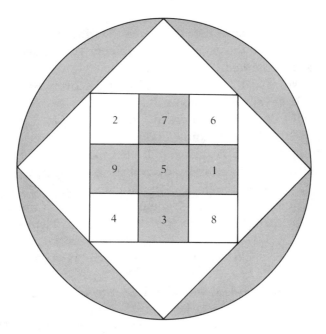

**Figure 140**  Dr. Matrix's lo shu belt buckle

"Yes. All even numbers are yin, all odd numbers yang. The central 5 is identified with the earth. Metal is indicated by 4 and 9, fire by 2 and 7, water by 6 and 1 and wood by 8 and 3. Each doublet has a yin-yang balance. The *lo shu* is the only way, as you know, aside from rotations and reflections, to arrange the nine digits so the sum of each row, column and main diagonal is the same. I think you gave a simple proof of uniqueness in one of your columns."

I nodded. "In my January 1976 column on magic squares and cubes. The *lo shu* has always impressed me as one of the most elegant patterns in the history of combinatorial number theory."

"It has many astonishing properties," said Dr. Matrix, "only a few of which I recall your mentioning."

I whipped a notepad and pencil from a pocket.

"For example, imagine the *lo shu* on a torus, its surface divided into nine cells. If you prefer, think of the plane square of three-by-three cells as torus-connected. Each pair of adjacent cells in the top row is considered joined to the two cells on the bottom row directly beneath. Each pair of adjacent cells in the left column is considered attached to the corresponding pair in the right column."

"And," I interrupted, "the four corner cells are also joined."

"Precisely. Altogether the toroidal matrix has nine two-by-two squares."

"So?"

"The sums of the four digits in each of the nine squares not only are all different, they are the consecutive numbers 16 through 24."

"Weird!" I exclaimed. "I didn't know that."

"I'm not surprised," said Dr. Matrix. "I discovered it when I was seven, but I've never written about it."

I jotted all this down, then asked: "Does the *lo shu* have any other curious properties I should know about?"

"Hundreds," Dr. Matrix answered. "Let me describe one I discovered 30 years ago. It's been rediscovered since, though few mathematicians know it. Consider the three-digit numbers given by the three rows. Square each number and add the squares. Do this with the three numbers formed by taking the same triplets in reverse. The two sums are identical."

I took out my pocket calculator. Dr. Matrix was right:

$$276^2 + 951^2 + 438^2 = 672^2 + 159^2 + 834^2 = 1,172,421$$

"There is more," said Dr. Matrix, adjusting his monocle. "The same result holds for the three-digit numbers formed by the three columns."

"How about the diagonals?"

"*Oui, mon ami.* The main and broken diagonals have the same identities. Of course, they are all straight diagonals on the torus. Take the three diagonals that go down and right. Square each number and add. The sum equals the sum of the squares of the same three numbers reversed. This holds true also for the three diagonals that go down and left."

I verified this on my calculator:

$$258^2 + 714^2 + 693^2 = 852^2 + 417^2 + 396^2 = 1{,}056{,}609$$
$$654^2 + 798^2 + 213^2 = 456^2 + 897^2 + 312^2 = 1{,}109{,}889$$

Dr. Matrix pressed a concealed button at the base of his belt buckle, and the silver disk dropped into his palm. He turned it over and placed it beside my plate. Figure 141 (left) shows the square engraved on the back.

"You are looking," said Dr. Matrix, "at one of the most incredible magic squares ever discovered. It was found a few years ago by my friend Lee Sallows. He calls it the *li shu*."

"I can see it is magic," I said, "but it must have some other unusual property."

"It does indeed."

Dr. Matrix sketched an empty nine-cell matrix on my notepad. In each cell he put a number that counted the number of letters in the English word for each corresponding number on the *li shu*. Thus "five" has four letters, and so a 4 goes into the top left corner, and similarly for the other cells (Figure 141 right). I could hardly believe it. Not only was the new square magic, but its numbers were consecutive from 3 to 11!

Later I learned that "alphamagic" is Sallows' term for magic squares that translate in this preposterous way, in any given language, to another

| 5 | 22 | 18 |
|---|----|----|
| 28 | 15 | 2 |
| 12 | 8 | 25 |

| 4 | 9 | 8 |
|---|---|---|
| 11 | 7 | 3 |
| 6 | 5 | 10 |

**Figure 141**   *Lee Sallows' li shu* (left) *and its alphamagic partner* (right)

magic square. Results of his in-depth computer investigations of alpha-magics of all sizes and in more than 20 languages were reported in 1986 in his two-part article "Alphamagic Squares." Sallows is a British electronics engineer who works for the University of Nijmegen in Holland.

Iva had been listening with a faint smile. "Perhaps," she said to her father, "you can give Mr. Gardner some *lo shu*–related puzzles that would entertain his readers if he writes about you again."

Without hesitation Dr. Matrix proposed a dozen problems from which I selected the following:

1. Construct a three-by-three magic square using nine consecutive whole numbers (nonnegative integers) that has a constant of 666, the Bible's notorious "number of the beast."

2. Figure 142 (left) shows a matrix on which 8 is on a side instead of in a corner as on the *lo shu*. Add eight whole numbers to the vacant cells, no two of the nine numbers alike, to form a magic square with the *lo shu's* constant of 15.

3. Construct a magic square with nine prime numbers, no two alike, that has the lowest possible magic constant. The number 1 is excluded because today it is not considered a prime. The number 2 also is unusable. (Proof: Assume 2 appears on an *n*-by-*n* prime magic square, and that *n* is odd. Because 2 is the only even prime, all rows and diagonals would have an odd sum except those containing 2, which would be even. If *n* is even, all rows and diagonals would have an even sum except those containing 2, which would be odd.)

4. Adopting the simple code of $A = 1$, $B = 2$ and so on, translate the *lo shu* digits into the nine letters shown in Figure 142 (right). Imagine a chess king placed on a cell. Make seven

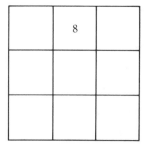

**Figure 142**   *Two lo shu puzzles*

king moves to spell a hyphenated eight-letter word, no two letters alike, counting the letter where the king starts as the word's first letter.

Dr. Matrix replaced the silver disk on his belt buckle. "By the way," he said, leaning back in his chair and adjusting his monocle, "the three horizontal numbers on the *lo shu*, as they appear on my *lo shu*'s orientation, are interesting. Take 276, the top number. It's the sum of the fifth powers of 1, 2 and 3. The middle number 951 is the difference between the squares of consecutive numbers 475 and 476; 475 is the difference between the squares of consecutive numbers 237 and 238; and 237 is the difference between the squares of consecutive numbers 118 and 119.

"My head spins," I said, as I jotted this down. "Is there anything remarkable about 438, the bottom number of your *lo shu*?"

"It is the most *remarquable* of all," said Dr. Matrix. "It's a Smith number."

I looked blank. "What the blazes is a Smith number?"

Smith numbers, Dr. Matrix explained, were discovered by Harold Smith, a brother-in-law of a mathematician at Lehigh University in Bethlehem, Pennsylvania, Albert Wilansky. One day Smith, who is not a mathematician, noticed that if his phone number were written as a single number 4,937,775, it had a whimsical property. Add all the digits in its prime factors 3, 5, 5, 65,837, and you get 42, which is also the sum of the digits in the phone number. Wilansky called any composite number with this property a Smith number and wrote a short note about it for the *Two-Year College Mathematics Journal*. Later Wayne McDaniel at the University of Missouri proved that there is an infinity of Smiths.

The smallest Smith number is 4. Among the first million integers there are 29,928 Smiths. This is almost 3 percent, and it is believed that this percentage holds approximately for any interval of a million integers. The chart in Figure 143 lists the Smiths less than 2000. Note that 666 is a Smith number, as is 1776. Dr. Matrix called my attention to the curious coincidence that 1776 is both the publication date of Adam Smith's *The Wealth of Nations*, the bible of capitalism, and the date of the founding of the world's most capitalist nation.

Dr. Matrix digressed to explain a curious property of 1776 that he had discovered. He asked me to select any digit $N$ and then punch the $N$ key of my calculator three times to put $NNN$ in the readout. I was asked to multiply $NNN$ by 16 and then divide the product by $N$. Result: 1776!

Two consecutive Smiths, such as 728 and 729 (the next highest pair is 2964 and 2965), are called Smith brothers. It is not known if the

| | | |
|---|---|---|
| 4 | 645 | 1507 |
| 22 | 648 | 1581 |
| 27 | 654 | 1626 |
| 58 | 663 | 1633 |
| 85 | 666 | 1642 |
| 94 | 690 | 1678 |
| 121 | 706 | 1736 |
| 166 | 728 | 1755 |
| 202 | 729 | 1776 |
| 265 | 762 | 1795 |
| 274 | 778 | 1822 |
| 319 | 825 | 1842 |
| 346 | 852 | 1858 |
| 355 | 861 | 1872 |
| 378 | 895 | 1881 |
| 382 | 913 | 1894 |
| 391 | 915 | 1903 |
| 438 | 922 | 1908 |
| 454 | 958 | 1921 |
| 483 | 985 | 1935 |
| 517 | 1086 | 1952 |
| 526 | 1111 | 1962 |
| 535 | 1165 | 1966 |
| 562 | 1219 | |
| 576 | 1255 | |
| 588 | 1282 | |
| 627 | 1284 | |
| 634 | 1376 | |
| 636 | 1449 | |

**Figure 143** *Smith numbers to 2000.*

number of Smith brothers is infinite. It has been shown, however, that there is an infinity of what Dr. Matrix called Psmiths (Iva preferred the term Smithtims)—palindromic Smiths that read the same in both directions.

Figure 144 copies a sketch by Dr. Matrix of a magic square made with nine Smiths. The constant 822, he assured me, was the lowest possible for such a square. Halve each number in this square, and you get a square made with nine different primes and a constant of 411.

Dr. Matrix reminded me that 22, the second Smith, is the number on which Rick, in the movie *Casablanca*, advised a patron in dire financial need to place his bets at the roulette table. The gaffed wheel stopped three times on 22. The fifth Smith, 85, Dr. Matrix added, is the sum of the letters in MATRIX, using the $A + 1$, $B = 2$ code.

Dr. Matrix referred to his friend Samuel Yates, a retired computer scientist now living in Delray Beach, Florida, as the world's top expert

| 94 | 382 | 346 |
|----|-----|-----|
| 526 | 274 | 22 |
| 202 | 166 | 454 |

C = 822

**Figure 144**   *The lowest three-by-three Smith magic square*

on Smith numbers. Smiths are related to what are called repunits — numbers consisting entirely of 1's. From every repunit whose prime factors are known one can construct a Smith number. Since repunits are infinite, the infinity of Smiths follows. Yates has long been an authority on repunits, having published many papers and one entire book about them.

Only five repunit primes are known: $R_2$, $R_{19}$, $R_{23}$, $R_{317}$ and $R_{1031}$. The subscripts, which also must be prime, indicate the number of 1's in the repunit prime. In 1987 Yates obtained from $R_{1031}$ the largest known Smith, a number of 10,694,985 digits. It is the product of

$$9 \times R_{1031} \ (10^{4594} + 3 \times 10^{2297} + 1)^{1476} \times 10^{3913210}$$

"Are there other Smiths on the *lo shu*?" I asked.

"Incredibly, yes," Dr. Matrix replied. "The two diagonals, taking the digits right to left, are both Smiths. The prime factors of 654 (2, 3 and 109) have digits that add to 15, and the same is true of 852 and its factors 2, 2, 3 and 71."

"Do the vertical numbers of the *lo shu* have any strange properties?"

"My dear Gardner," Dr. Matrix said in a patronizing tone, "*no* number exists that does not have strange properties. Consider, for instance, your hotel room in Lisbon. Iva told me its number was 243. A most unusual number. Because its prime factors are 3, 3, 3, 3 and 3, it is written as 100000 in ternary notation. Do you know what its reciprocal is in decimal notation?"

I shook my head.

Dr. Matrix put on his monocle to write on my notepad:

$$1/243 = 0.004115226337448559 \ . \ . \ .$$

"Dick Feynman told me about that crazy fraction when I was at Los Alamos during the Second World War to check some numbers involved in the construction of the atom bomb."[*]

"Did you know Feynman before then?"

"His teacher," replied Dr. Matrix, "was my pupil. By the way, are you aware that *flatcar* is an anagram of *fractal*? But I digress. Every three-digit number on the *lo shu*, taking any row, column or diagonal in either direction, has the following extraordinary property. Add the number to itself, multiply it by itself, subtract it from itself and divide it by itself. When you add the four results, you always get a perfect square."

I selected 654 at random. Sure enough, my calculator showed that $(654 + 654) + (654 - 654) + (654 \times 654) + (654 \div 654) = 429{,}025$, the square of 655. It was not until several days later that I realized how I had been flimflammed. The property holds for any number whatever. Here's a simple algebraic proof:

$$(n + n) + (n - n) + (n \times n) + (n \div n) =$$
$$2n + 0 + n^2 + 1 =$$
$$n^2 + 2n + 1 =$$
$$(n + 1)^2$$

I don't want to give the impression that Iva was left out of our conversation. The three of us talked about many things during dinner and later in the gambling area, where Dr. Matrix won a sizable amount of money at *vingt-et-un* (twenty-one), as the French call what American casinos call blackjack. Here I have given space only to material related to word and number play.

I aimed my pencil at Dr. Matrix's silver belt buckle. "That circle and the two inscribed squares intrigue me."

"Good you mentioned it," said Dr. Matrix. "Your fans might enjoy an amusing puzzle based on the design. Assume the circle's radius is 7 units. How quickly can you determine the length of the side of the smaller square?"

---

[*]The fraction is mentioned on page 116 of Richard Feynman's autobiography, *Surely You're Joking, Mr. Feynman* (Norton, 1985). "It goes a little cockeyed after 559," Feynman comments, ". . . but it soon straightens itself out and repeats itself nicely." Feynman recalls writing about the fraction in a letter when he was working at Los Alamos. The project's censor returned the letter because he thought the number might be a code. Feynman sent the censor a note explaining that it couldn't be a code "because if you actually *do* divide 1 by 243, you do, in fact, *get* all that, and therefore there's no more information in the number 0.004115226337 . . . than there is in the number 243 — which is hardly any information at all."

"Hmmm. I'll have to do some calculations based on Euclid's Pythagorean theorem."

"*Au contraire,*" Dr. Matrix said while Iva grinned. "The answer is obvious."

We had just been served our after-dinner wine. To cover my *embarras* I touched my glass to Dr. Matrix's glass. "I know that the problems of three little people don't amount to a hill of beans," I said, trying to sound like Bogart, "nevertheless, permit me, Mr. Lundy, to wish you a long and happy revivification."

An elderly black man at the piano was playing "As Time Goes By." Iva, simulating a look of pain, clinked her glass against mine and said, in a much better imitation of Bogie, "Here's looking at *you*, clid."

## ANSWERS

1.  The "beast" square (Figure 145, top left) is obtained by adding $(666 - 15)/3 = 217$ to each cell of the
    *lo shu.*

| 219 | 224 | 223 |
|-----|-----|-----|
| 226 | 222 | 218 |
| 221 | 220 | 225 |

| 1  | 8 | 6 |
|----|---|---|
| 10 | 5 | 0 |
| 4  | 2 | 9 |

| 17 | 113 | 47  |
|----|-----|-----|
| 89 | 59  | 29  |
| 71 | 5   | 101 |

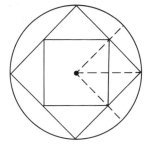

***Figure 145*** *Solutions to problems*

2. The only solution is shown at top right, Figure 145.

3. The unique solution is at bottom left, Figure 145. The constant is 177.

4. The word is "big-faced."

The question about the geometrical design on Dr. Matrix's belt buckle is answered as follows. The side of the smaller square is 7, the same as the circle's radius. The bottom right drawing in Figure 145 is the "look-see" proof.

## ADDENDUM

I find in my notes that Dr. Matrix had called my attention to the following connection between the *li shu* square and the "number of the beast." Add 100 to each cell, and do the same for its partner. When corresponding cells of the two squares are added, the result is a magic square with a constant of 666.

In working on problems involving three-by-three magic squares, the algebraic structure shown in Figure 146 is useful. Nine numbers will form a magic square if and only if they can be grouped in three sets of triplets, each triplet an arithmetic progression with the same common difference for all three progressions and with the smallest numbers of the triplets in another arithmetic progression. The triplets are distinguished in the illustration by different shades of gray.

Any real numbers can be substituted for $a$, $b$ and $c$. The central number $a$ is always one-third of the magic constant. Thus if the constant is pi, $a$ must be one-third of pi. Of course, proper values have to be chosen if you want a magic square with no duplicate numbers. If the numbers are to be consecutive positive integers, the constant must be a multiple of 3, $a$ must be 5 or greater, $b$ must be 1 and $c$ must be 3.

In 1987, in an article on prime magic squares for a 1988 *Mathematical Sciences Calendar* (Rome Press, 1987), I offered $100 to the first person who could construct a three-by-three magic square with consecutive primes. I repeated the challenge at the close of my *Riddles of the Sphinx* (Mathematical Association of America, 1987). Early in 1988 Harry Nelson, using a Cray computer at the University of California and an ingenious program, won the prize with 22 solutions of which the one with the lowest constant is shown in Figure 147. The program did not prove that this is the lowest possible, although the probability is almost certain that it is.

| | | |
|---|---|---|
| a − c | a − b + c | a + b |
| a + b + c | a | a − b − c |
| a − b | a + b − c | a + c |

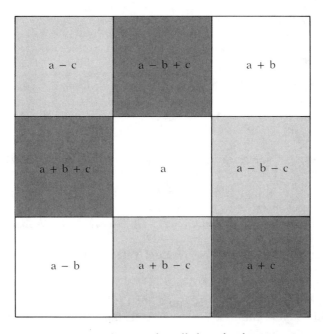

***Figure 146*** *Schemata for all three-by-three magic squares*

| | | |
|---|---|---|
| 1480028201 | 1480028129 | 1480028183 |
| 1480028153 | 1480028171 | 1480028189 |
| 1480028159 | 1480028213 | 1480028141 |

***Figure 147*** *A magic square with consecutive primes*

# BIBLIOGRAPHY

### On alphamagic squares

"Alphamagic Squares." Lee Sallows, in *Abacus*, 4, 1986, pp. 28–45 and 1987, pp. 20–29, 43.

### On Smith numbers

"Smith Numbers." Albert Wilansky, in *Two-Year College Mathematics Journal*, 13, 1987, p. 21.

"The Existence of Infinitely Many *k*-Smith Numbers." W. L. McDaniel, in *The Fibonacci Quarterly*, 56, 1983, pp. 36–37.

"Construction of Smith Numbers." Sham Oltikar and Keith Weiland, in *Mathematics Magazine*, 56, 1983, pp. 36–37.

"Special Sets of Smith Numbers." Samuel Yates, in *Mathematics Magazine*, 59, 1986, pp. 293–296.

"Smith Numbers Congruent to 4 (Mod 9)." Samuel Yates, in *Journal of Recreational Mathematics*, 19, 1987, pp. 139–141.

"Palindromic Smith Numbers." Wayne McDaniel, in *Journal of Recreational Mathematics*, Vol. 19, 1987, pp. 34–37.

"How Odd the Smiths Are." Samuel Yates, in *Journal of Recreational Mathematics*, Vol. 19, 1987, pp. 268–74.

"Review of Smith Numbers." Donald Bushaw, in *The College Mathematics Journal*, Vol. 19, 1988, p. 375.

"The Sum of Digits Function and its Application to a Generalization of the Smith Numbers Problem." Wayne McDonald and Samuel Yates, in *Nieuw Arch ief voor Wiskunde, Vierde Serie*, Vol. 7, Nos. 1–2, March/July, 1989, pp. 39–51.

"How to Generate All Types of Smith Numbers." Robert Bishop, in *Journal of Recreational Mathematics*, Vol. 22, 1990, pp. 262–270.

"In Search of Perfect Smiths." James L. James, in *Journal of Recreational Mathematics*, Vol. 25, No. 2, 1993, pp. 109–117.

"Smith Numbers." Underwood Dudley, in *Mathematics Magazine*, Vol. 67, February 1994, pp. 62–65. See also Wayne McDaniel's letter, ibid., Vol. 57, October 1994, pp. 315–317.

# *Postscript*

Penrose tiles are now commercially available. From Kadon Enterprises, 1227 Lorene Drive, Pasadena, Maryland, 21122, you can purchase a set of darts and kites, a set of fat and thin diamonds, and a set of two kinds of birds, based on Penrose tiling, that are called "Perplexing Poultry." The birds are also available in England from Pentaplex, Ltd., Royal House, Brighouse, West Yorkshire, HD6 ILQ.

In his 1994 book *Shadows of the Mind* (page 32) Roger Penrose reports on his discovery, based on tiles found by Robert Ammann, of three polyominoes that tile the plane only nonperiodically. They are shown in Figure 1. Penrose also reveals an asymmetric polyomino he found that tiles periodically, but only when all eight of its orientations are used.

Penrose became famous among nonmathematicians and nonphysicists in 1989 when Oxford University Press published his *Emperor's New Mind* and the book became a best seller. It is a smashing attack on AI (artificial intelligence) enthusiasts who think that very soon computers will cross

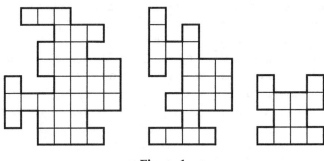

*Figure 1*

a threshold of complexity, and become self aware, with minds capable of doing everything a human can do. I had the honor of writing the book's foreword.

Penrose followed the book in 1994 with *Shadows of the Mind*, also published by Oxford. It defends his view in greater depth. Penrose does not deny that the mind is a function of the brain, but he believes that our brains operate in ways not yet understood. No computer, in his opinion, will ever become conscious of what it is doing so long as it operates solely by algorithms that tell switches how to divert electrical currents here and there. In brief, computers of the sort we know how to build, differ in principle only from mechanical calculating machines in the speed by which they twiddle symbols. Penrose is convinced that no computer will rival human intelligence until we know more about how quantum mechanics, and perhaps phenomena deeper than quantum mechanics, are used by our mysterious brains. Both his books have been severely savaged by AI leaders, but this is a long story, too remote from mathematics to discuss further.

On pages 21–22 I gave the misleading impression that the continued fraction was a discovery of Conway's. Jeffrey Shallit imformed me that it was known at least as early as 1926 when P. E. Böhmer discussed it in a German paper. See J. L. Davison, "A Series and Its Associated Continued Fraction," in the *Proceedings of the American Mathematical Society*, vol. 65 (1977), pages 194–98.

Benoit Mandelbrot has become increasingly famous, with many new honors given to him, for his work on fractals. Since I wrote about fractals, the new field of chaos theory, intimately related to fractals, has exploded on the mathematical scene. In the chapter's bibliography I have tried to select the most accessible books on chaos and fractals out of an enormous, rapidly proliferating literature.

Martin Kruskal, now a mathematician at Rutgers University, became fascinated by Conway's surreal numbers. For many years he has been working on their clarification and elaboration, and their potential applications to other fields of mathematics. For an introduction to this exciting work, see the last two entries, by Polly Shulman and Robert Matthews, in the chapter's bibliography under the subhead of "Surreal Numbers."

Will Shortz, now crossword puzzle editor of *The New York Times*, is planning a book about Sam Loyd and the history of his puzzles. It will contain hundreds of Loyd puzzles not in the famous *Cyclopedia*. Shortz sent me a copy of an 1899 trade card advertising Pond's Extract, a patent medicine pain killer. It contains Loyd's Klondike puzzle with the two digits altered. Loyd correctly announces a solution in nine steps. The "troublesome 2″ remains on the card. Loyd's solution is indeed "best" if that means shortest, but he was wrong in the *Cyclopedia* to call it unique.

David Fabian, of Houston, tackled the problem of searching for the shortest possible Chinese checkers game on boards with ten or fifteen marbles per side. Using the notation in Figure 2, he found these solutions:

| FIFTEEN MARBLES | | | | TEN MARBLES | |
|---|---|---|---|---|---|
| 1. C7–C6 | F1–F3 | 10. E8–E2 | E1–C7 | 1. C8–D8 | H2–H4 |
| 2. A5–C5 | F2–F4 | 11. B9–D3 | H1–B9 | 2. D9–D7 | I4–G4 |
| 3. C5–D5 | I4–E4 | 12. B7–H3 | F3–B7 | 3. A7–E7 | G1–G5 |
| 4. C9–G5 | G3–A5 | 13. C8–C4 | D1–D9 | 4. A8–E6 | H3–F3 |
| 5. E9–E8 | I3–C9 | 14. A6–A4 | G2–A6 | 5. E6–F6 | H1–E3 |
| 6. A7–E7 | I1–E9 | 15. A8–G2 | I2–A8 | 6. D7–C6 | G4–A8 |
| 7. A9–I3 | G1–A9 | 16. D8–D7 | F4–D8 | 7. A6–E4 | I2–A6 |
| 8. D9–F7 | H3–D1 | 17. B6–F4 | E4–C8 | 8. A9–C5 | F1–D9 |
| 9. F7–F6 | I5–A7 | 18. B8–F8 | H2–B8 | 9. C5–D5 | C3–A9 |
| | | 19. I3–I2 | H4–B6 | 10. B8–H2 | I3–A7 |
| | | | | 11. C9–G3 | H4–B8 |
| | | | | 12. B7–B6 | G2–C8 |
| | | | | 13. B9–F5 | G5–C9 |
| | | | | 14. F5–F4 | F3–B9 |
| | | | | 15. E4–D4 | H1–B7 |

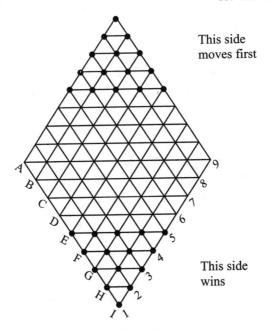

This side
moves first

This side
wins

*Figure 2*

Richard Guy and Patrick Kelly have conjectured that the no-3-in-line patterns shown in Figure 36, for orders 2, 4, and 10, are the only patterns with the symmetry of a square. This has been verified up to $n = 60$. Achim Flammenkamp, in a 1992 paper listed in the bibliography, gives solutions for orders 38, 40, 42, 44, and 46, and new solutions for many other orders.

G. M. Hamilton, I. T. Roberts, and D. G. Rogers consider the generalized pool-ball problem in depth in their paper "Regular Perfect Systems of Sets of Iterated Distances," A Curtin University of Technology Research Report, 1993. A sequel by the same three authors, "Uniform Perfect Systems of Sets of Iterated Differences of Size 4" appears in *Discrete Mathematics*, (volume 162, 1996, pages 133–150). and "The Problem of Irregular Perfect Systems of Sets of Iterated Distances," by Ramsey and Rogers, is in *Graphs and Combinatorics* (volume 13, 1997).

In discussing Scott Kim's tetrads I said it was not known if one existed with polygons of order less than 12. Walter Trump, of West Germany, answered yes with the order-11 polyomino shown in Figure 3. Trump also sent the tetrad of congruent hexagons (Figure 4) which has the property that any pair of hexagons will form a square.

Figure 5 shows Trump's two-hole tetrad made with polygons that are bilaterally symmetric in two directions, and having a border with the symmetry of an equilateral triangle. Figure 6 shows his tetrad without holes, made of bilaterally symmetric polygons. This pattern was independently sent by Karl Scherer, also of West Germany.

The paper by Chung, Szemerádi, and Trotter, cited on page 136, appeared in *Discrete and Computational Geometry*, volume 7, number 2, 1992. Michael Beeler's computer results, mentioned on page 126, were published in the *Journal of Recreational Mathematics*, vol. 10, number 1, 1977–78. See also "Distinct Distances Determined by Subsets of a Point Set in Space," by David Avis, Paul Erdös, and János Pach, in *Computational Geometry I* (Elsevier, 1991).

Richard Feynman's paper on negative probability was published in *Quantum Implications: Essays in Honour of David Bohm*, edited by B. J. Hiley and F. David Peat, Routledge and Kegan Paul, London, 1987, pages 235–248. An example of the use of negative probability is given in *Concrete Mathematics*, second edition (1994), by Ronald Graham, Donald Knuth, and Oren Patashnik (Addison-Wesley Publishing Company).

Surprising developments in trapdoor ciphers have taken place since I broke the news about them in my August 1977 *Scientific American* column. In 1990 a 155-digit composite number was factored into its three primes by a team headed by Arjen Lenstra, of Bellcore, and Mark Manasse, of Digital Equipment Corporation. The task required the use of a thousand computers working together over several months. This did not, of course, render the MIT code, now known as the RSA system after the initials of its three formulators, but it does force the many users of the system to base it on larger composite numbers, say of 200 digits.

Four years later, in 1994, Lenstra did it again. The message I had published as a challenge, with $100 to the first to crack it, was solved by Lenstra and his associates. The message proved to be "The magic words are squeamish ossifrage." Rivest had selected the last two words at random, and soon forgot what they were. His $100 check went to Lenstra. Breaking the ciphertext required factoring of a 129-digit composite number known as RSA-129. The task required some 600 persons using 1,600 computers, linked by the Internet, that worked on the factoring for eight months.

The following references deal with Lenstra's remarkable achievement:

"To Break the Unbreakable Number," William Booth, in *The Washington Post*, June 25, 1990, page A3.

"The Assault ... ." Gina Kolata, in *The New York Times*, March 22, 1994.

"Small Army of Code-Breakers Conquers a 129-digit Giant." Gary Taubes, in *Science*, vol. 264, May 6, 1994, pages 776–77.

"Team Sieving Cracks a Huge Number." Ivars Peterson, in *Science News*, vol. 14, May 7, 1994, page 292.

"The Magic Words are Squeamish Ossifrage." Brian Hayes, in *American Scientist*, vol. 8, July/August 1994, pages 312–314.

"Superhack." Kristin Leutwyler, in *Scientific American*, July 1994, pages 17–20.

"The Magic Words are Squeamish Ossifrage." Barry Cipra, in *SIAM News*, vol. 27, July 1994.

"Wisecrackers." Steven Levy, in *Wired*, March 1996, page 129ff.

Two recent papers on how to play games over the phone using such randomizers as coin flips, dice, and cards, are: "Flipping a Coin Over the Telephone," by Charles Vanden Eynden, in *Mathematics Magazine*, vol. 62, June 1989, pages 167–72, and "Mathematical Entertainments: Playing Games Over the Telephone," by David Gale, in *The Mathematical Intelligencer*, vol. 14, no. 3, 1992, pages 60–64.

Later references on zero-knowledge proofs include: "Zero Knowledge Proofs," by Catherine McGeoch, in *The American Mathematical Monthly*, August-September 1993, pages 682–685; and "Proof of Purchase on the Internet," by Ian Stewart, in *Scientific American*, February 1996, pages 124–125.

Another astonishing development has been the discovery that unbreakable codes are theoretically possible using computers that operate on quantum mechanical principles. No such computers are likely to be built in the near future, but their possibility is currently generating a large number of technical papers by Charles Bennett, of IBM, and others.

The discovery that Ramsey Number $R(4,5)$ equals 25 was first announced in 1993 by the *Rochester Democrat and Chronicle* because one of the discoverers, Stanislaw Radziszowski, was a professor at the Rochester Institute of Technology. Someone using the initials B.V.B. reported this to Ann Landers who published the letter in her syndicated advice column on June 22, 1993. B.V.B. put it this way:

> Two professors, one from Rochester, the other from Australia, have worked for three years, used 110 computers and communicated 10,000 miles by electronic mail, and finally have learned the answer to a question that has baffled scientists for 63 years. The question is this: If you are having a party and want to invite at least four people who know each other and five who don't, how many people should you invite? The answer is 25. Mathematicians and scientists in countries worldwide have sent messages of congratulations.

> I don't want to take anything away from this spectacular achievement, but it seems to me that the time and money spent on this project could have been better used had they put it toward finding ways to get food to the millions of starving children in war-torn countries around the world.

Miss Landers replied as follows:

> Dear B.V.B.: There has to be more to this "discovery" than you recounted. The principle must be one that can be applied to solve important scientific problems. If anyone in my reading audience can provide an explanation in language a lay person can understand, I will print it. Meanwhile I am "Baffled in Chicago."

I sent Miss Landers this letter:

> You asked (June 22) for an explanation of the proof that 25 people are required at a party if at least four are mutually acquainted, and at least five are mutual strangers. This is a theorem in what is called Ramsey theory. It can be modeled by points on paper, each point representing a person. Every point is joined to every other point by a line.

> If a line is colored red it means that its two points are persons who know each other. If a line is blue, it means its two persons are strangers. We now can ask: what is the smallest number of points needed so that no matter how the lines are bicolored, there

is sure to be either four points mutually joined by red lines, or five mutually joined by blue lines? The answer, 25, was not proved until recently. For the problem's background, see the chapter on Ramsey theory in my book *Penrose Tiles to Trapdoor Ciphers* (W. H. Freeman, 1989).

As far as I know, the letter was not printed. Perhaps Miss Landers remained baffled as to what earthly use this result could possibly have.

Radziszowski and his associate Brendon McKay published their proof in a paper titled "$R(4,5) = 25$" in *The Journal of Graph Theory*, vol. 19, no. 3, 1995, pages 309–321.

On page 173 I said that I knew of no simple way to determine how the area of a 3, 4, 5, right triangle can be quartered by two perpendicular lines. Sin Hitotumatu showed how to do it in his paper "On a Quadrisection Problem of M. Gardner," in *Research Activities*, Vol. 4, 1992, pages 1–4, published in English by the Faculty of Science and Engineering, Tokyo Denki University.

# Index